ATOMIC PHYSICS OF LASERS

ATOMIC PHYSICS OF LASERS

D.A. Eastham

SERC, Daresbury Laboratory, Warrington, England

Taylor & Francis
London and Philadelphia
1986

UK Taylor & Francis Ltd, 4 John St, London WC1N 2ET

USA Taylor & Francis Inc., 242 Cherry St, Philadelphia, PA 19106–1906

British Library Cataloguing in Publication Data

Eastham, D. A.
Atomic physics of lasers.
1. Lasers
I. Title
535.5′8 QC688

ISBN 0-85066-343-1

Library of Congress Cataloging-in-Publication Data

Eastham, D. A. (Derek A.)
Atomic physics of lasers.

Bibliography: p.
Includes index.
1. Lasers. I. Title.
QC688.E37 1986 535.5′8 85-30356
ISBN 0-85066-343-1

Typeset by Mid-County Press, 2a Merivale Road, London
Printed in Great Britain by Taylor & Francis (Printers) Ltd, Basingstoke, Hants.

PREFACE

This monograph is intended to be used as an introduction to lasers, suitable for a 2nd or 3rd year undergraduate who has some basic groundwork in quantum mechanics and optics. I have attempted to make the book as self contained as possible. For this reason, only a few references are given, although a short bibliography is provided with each chapter.

The book is in no way an attempt at a comprehensive treatment but rather a gentle introduction which, hopefully, will provide the reader with enough basic concepts so that research papers, or more difficult texts, could be tackled. In this respect it should prove particularly useful to the new research worker who is starting to use lasers, or is beginning laser research.

Chapter 1 is a short review of the structure of atoms and molecules, whilst Chapter 2 deals with the semiclassical theory of the interaction of radiation and atoms. Chapter 3 provides the basic framework of laser physics and Chapter 4 considers some specialised topics. Descriptions of laser systems are given in Chapters 5 and 6.

I would like to thank all my colleagues for many helpful suggestions and discussions. In particular, I would like to thank Dr. J.A.R. Griffith for numerous suggestions and Dr. P.L. Knight for his (welcome) critical comments on the manuscript. Finally, I would like to thank Mrs. Sue Waller for carefully typing the manuscript.

Derek Eastham
Warrington
December, 1985

To Margriet, Paul and James

CONTENTS

CHAPTER 1

introduction and elementary review of the structure of atoms and molecules

1.0. Introduction

Lasers have assumed a unique, almost magical, place in the history of science and technology. From their first conception in 1955 they have found diverse applications in pure and applied science, and this has been accompanied, or sometimes preceded, by a rapid development in laser technology. It is the ability of lasers to produce an extremely intense, monochromatic beam of electromagnetic radiation, unlike any other source, which has given rise to this rapid expansion. For example, a small carbon dioxide laser, now manufactured with technology not much more advanced than that used to manufacture electric light bulbs, can produce a radiance which is 10^5 times greater than the sun. Applications of this high radiance range from cutting and welding of metals to the production of intense plasmas for thermonuclear fusion.

In other fields, particularly spectroscopy, it is the concentration of this radiation into a narrow frequency range (high spectral intensity), as well as the ability to tune the wavelength which are the most important. Thus, laser light can be used as an ultra-sensitive probe to selectively excite discrete energy levels in atoms and molecules, and because of the high intensity it is possible to study processes, like two photon absorption, which have very small cross sections. One of the most beautiful examples of this type of spectroscopy is the selective ionization of atoms or molecules using two or more tunable light beams. With this technique single atoms of one particular species can be detected! Another example is the production and study of planetary or Rydberg atoms; excited states of an atom where one or more electron is promoted to a very large radius orbit which can be approximately described using classical equations of motion. These atoms, which can have sizes as much as 10^3 times greater than a normal atom, cannot be produced by photon absorption using conventional sources but are quite easily formed using tunable laser beams. Tunable lasers are also exploited in resonance fluorescence spectroscopy. Here light is selectively absorbed when the laser frequency matches an atomic or molecular transition, and the spontaneous emission is used to signal the resonance position. Because of the extreme sensitivity, measurements are possible with very small quantities of atoms, and this can be exploited in many areas. For example, Figure 1.1 shows an

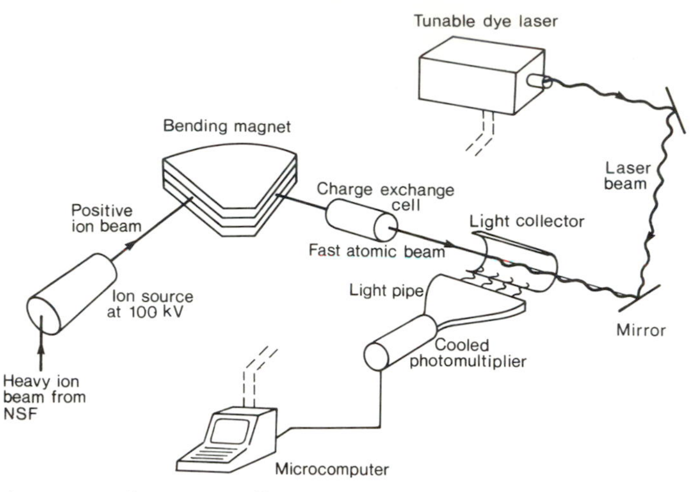

Figure 1.1. Arrangement for resonance fluorescence spectroscopy by colliding a laser beam with a fast atomic (ionic) beam from an isotope separator. This system is used to study hyperfine structure of radioactive atoms.

arrangement for measuring the hyperfine structure and isotope shifts of radioactive atoms. These measurements can be used to provide information about nuclear moments and radii.

In another sense lasers are unique. Nowhere else does one subject combine so many separate fields of study, from the details of quantum electrodynamics to theories of gaseous discharges. A list of major subjects would have to include quantum mechanics, atomic and molecular physics, optics and solid state physics, and a study of lasers might be approached from any of these viewpoints. However, it is the quality of combining and illustrating many of the basic concepts of physics which makes the study of lasers so attractive and exciting. Of course, the type of application, or laser, may influence the choice of approach, but, since the underlying principles by which all lasers operate is the same, a unified picture can be given. In this text the starting point is an elementary review of the theory of atomic and molecular structure.

The study of atomic structure is intimately related to the way atoms interact with electromagnetic radiation, and there are obvious advantages in relating the decay modes of excited atomic states to their structure. However, in this monograph, the two are separated and the detailed theory of the interaction of radiation is deferred until Chapter 2. The reader should be aware that this is a somewhat artificial separation, introduced in order to emphasize the most important aspects of laser physics.

1.1. Properties of atoms

Almost all of the chemical properties and most of the physical properties of atoms, except at extremes of temperature and pressure, are determined by the structure of the electronic cloud surrounding the nucleus. This structure is in turn directly related to the number of electrons in the atom, or the atomic number. Changes in neutron number for a given element produce only small changes in atomic structure and, apart from the obvious changes in density, do not significantly alter the chemical or physical properties. One of the striking characteristics of atoms is the occurrence of elements with widely different atomic numbers which have similar chemical properties. This periodicity can also be seen in the physical characteristics, and is a reflection of the recurrence of similar electronic structures.

Another property of atoms, indeed the most important as far as the generation of laser light is concerned, arises from the quantization of the energy of the atom. Atoms can only exist in definite discrete energy states and transitions between these can take place by the absorption or emission of electromagnetic radiation. In the decay of an atom from an excited energy level E_2, to a lower state E_1, the frequency of the photon, ν, is given by

$$E_2 - E_1 = h\nu \tag{1.1}$$

where h is the Planck constant. This equation also applies to the absorption of radiation. Thus, for example, the existence of discrete lines in the emission spectra of gaseous discharges or stellar objects is a manifestation of these quantized energy levels in atoms and ions. Figure 1.2 shows an example of how it is possible to use these to identify different atoms in remote parts of the universe.

The existence of these discrete atomic states can be related to the details of atomic structure, and to understand the nature of these it is necessary to formulate and solve the equations which determine the motion of electrons in atoms. We begin by studying the simplest atomic systems using the equations of quantum mechanics. The solutions to these are then used to illustrate the general properties of more complicated atoms where analytical solutions are not possible.

1.2. One electron atom and ions

The simplest systems to consider are those with a single electron orbiting the nucleus. This is the case of the hydrogen atom and the ions He^+, Li^{++}, Be^{+++},... where the number of crosses indicates the total ionic charge. In spectroscopy these are often designated HeII, LiIII, BeIV....

The Schrödinger equation for stationary states, ψ, of an electron bound in a Coulomb potential is written

$$\left(\frac{-\hbar^2}{2\mu}\nabla^2 + V(r)\right)\psi = E\psi \tag{1.2}$$

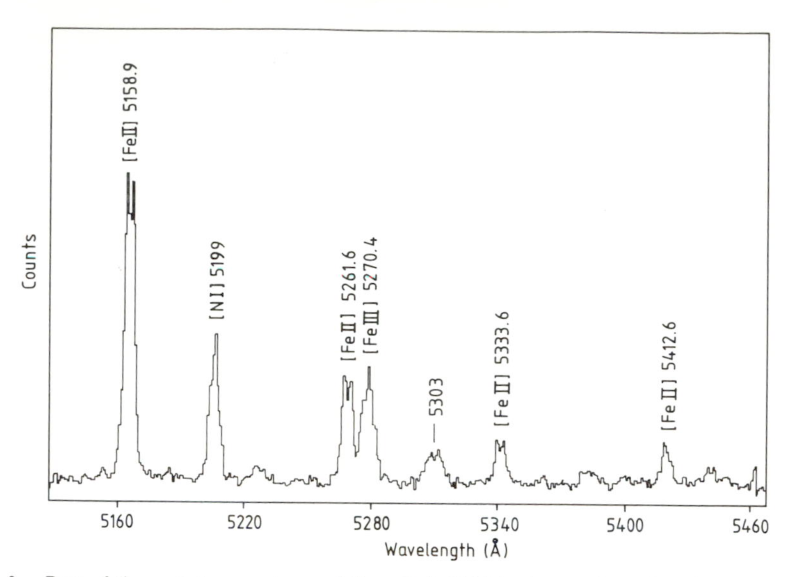

Figure 1.2. Part of the emission spectrum of the nebula N49 in the nearby galaxy known as the Large Magellanic cloud showing forbidden lines in Fe II and N I. The lines, whose rest wavelengths are marked, are shifted by about 6 Å due to the recessive motion (red shift) of the galaxy containing the nebula. Broadening of the lines is caused by turbulent and expansion motions in the nebula. The feature marked at 5303 is from Fe XIV and arises from gas at a temperature of 2×10^6 K, which is sufficiently hot enough to ionize iron atoms 13 times!
(Figure provided by Dr. Paul Murdin, Royal Greenwich Observatory.)

where

$$V(r) = \frac{-Ze^2}{4\pi\varepsilon_0 r}$$

and

$$\nabla^2 = \frac{1}{r^2}\left\{\frac{\partial}{\partial r}\left(r^2\frac{\partial}{\partial r}\right)+\frac{1}{\sin\theta}\frac{\partial}{\partial\theta}\left(\sin\theta\frac{\partial}{\partial\theta}\right)+\frac{1}{\sin^2\theta}\frac{\partial^2}{\partial\phi^2}\right\}$$

is the form for the Laplacian operator in spherical polar coordinates r, θ, ϕ. $\psi(r, \theta, \phi)$ is the wave function, μ is the reduced mass of the electron and E is the energy. Note that this equation is not relativistic and also does not include the interaction of the electron with the magnetic dipole field of the nucleus. A general solution to the differential Equation (1.2) is

$$\Psi_{nlm}(r, \theta, \phi) = R_{nl}(r)Y_{lm}(\theta, \phi) \tag{1.3}$$

which consists of a product of a spherical harmonic function $Y_{lm}(\theta, \phi)$ and a radial function $R_{nl}(r)$. Both of these functions are solutions of separate, though not independent, differential equations. The radial wave functions are given by the expression

$$R_{nl}(r) = -\left\{\left(\frac{2Z}{na_0}\right)^3 \frac{(n-l-1)!}{2n((n+l)!)^3}\right\}^{1/2} \exp(-\tfrac{1}{2}\rho)\rho^l L_{n+l}^{2l+1}(\rho) \qquad (1.4)$$

with $a_0 = 4\pi\varepsilon_0\hbar^2/\mu e^2$, $\rho = (2Z/na_0)r$, and L is the Laguerre polynomial. From the separated radial differential equation we can extract directly the energies of the states. These are given in terms of a positive integer n, known as the principal quantum number by the equation

$$E_n = -\frac{1}{(4\pi\varepsilon_0)^2}\frac{Z^2 e^4 \mu}{2\hbar^2 n^2} \qquad (1.5)$$

which describes the gross structure of the energy levels in single electron systems.

The spherical harmonics are solutions to the equation

$$\mathbf{l}^2 Y_{lm}(\theta, \phi) = l(l+1)\hbar^2 Y_{lm}(\theta, \phi) \qquad (1.6)$$

where $\mathbf{l}^2$ is given explicitly by the relation

$$\mathbf{l}^2 = -\frac{1}{\sin\theta}\left(\frac{\partial}{\partial\theta}\sin\theta\frac{\partial}{\partial\theta}\right) - \frac{1}{\sin^2\theta}\frac{\partial^2}{\partial\theta^2}$$

Evidently, $\mathbf{l}^2$ is identified with the orbital angular momentum of the electron and Equation (1.6) shows that the angular momentum of a particular state is quantized in units of $\hbar$, and has a magnitude $\sqrt{l(l+1)}\,\hbar$. All positive integer values of the orbital angular momentum quantum number, l, are allowed up to a maximum value of $n-1$. The magnitude of the angular momentum along the quantum axis (this can be any defined axis in spherically symmetric systems) is $m\hbar$, where m is known as the orbital magnetic quantum number, and can have all integer values from $-l$ to $+l$.

The solutions to the single electron system as determined by Equation (1.2) are therefore many fold degenerate; that is for each energy given by Equation (1.5) there are n^2 different solutions which have the same energy. Some of the l degeneracy is removed when the atom is considered relativistically, but the m degeneracy is never removed even when more complex atoms are considered. This arises because atoms are spherically symmetric and there can be no preferred quantum axis except when external fields are applied. (Hence the term magnetic quantum number.) All m values are equally probable and the electron cloud surrounding the nucleus has a spherical shape. Mathematically this means that the expression for the electron 'density' $\Psi^*\Psi$, where Ψ^* is the complex conjugate of Ψ, must be independent of θ and ϕ when summed over all values of m. This can easily be shown to be true for the single electron case using the properties of the spherical harmonics.

1.3. Fine structure

For more accurate values of the energies of the bound states the problem needs to be considered relativistically. This can be done rigorously using the relativistic Dirac equation, but fairly accurate values, particularly for the case of the hydrogen atom, can be obtained by considering correction terms to the basic equation, and using the results of first order perturbation theory. The first correction which can be considered is the change of the electron mass with velocity which of course is ignored in Equation (1.2). This produces a correction which becomes smaller as n is increased and the electron velocity is reduced. It can often be neglected when considering the excited levels of heavier atoms. A second correction to the levels is caused by the spin of the electron. In the Dirac equation this spin arises as a natural consequence of the dual nature of the solutions to the equation. However, this energy correction can be considered in a semi-classical way by assuming that the electron has a magnetic dipole moment, $\boldsymbol{\mu}$, which is related to the electron spin vector by the equation

$$\boldsymbol{\mu} = -\frac{2\mu_B \boldsymbol{s}}{\hbar} \tag{1.7}$$

where μ_B is the Bohr magneton and is equal to $e\hbar/2m$. $\boldsymbol{s}$ is the electron spin vector and is defined in the operator equations

$$\boldsymbol{s}^2\chi = s(s+1)\hbar^2\chi \tag{1.8}$$

and

$$s_z\chi = m_z\hbar\chi$$

where the electron spin quantum number s is $\frac{1}{2}$ and m_z, its projection on the quantum axis, can have values $\pm\frac{1}{2}$. χ is the spin wave function. The dipole moment of the electron can then interact with the magnetic field generated by the orbital motion about the nucleus to give an energy shift of $-\boldsymbol{B}.\boldsymbol{\mu}$, where $\boldsymbol{B}$ is the magnetic field. $\boldsymbol{B}$ depends on the orbital angular momentum of the electron and the full expression for the energy shift is

$$\Delta E = \frac{-\alpha^2 Z^2}{n^2\hbar^2} E_n \frac{n}{l(l+\frac{1}{2})(l+1)} \boldsymbol{l}.\boldsymbol{s} \tag{1.9}$$

where α is the fine structure constant and E_n is the unperturbed energy given by Equation (1.2). α is a dimensionless constant given by the equation

$$\alpha = \frac{e^2}{\hbar c}\left(\frac{1}{4\pi\varepsilon_0}\right) \tag{1.10}$$

and has an approximate value of 1/137. Equation (1.9) defines the energy due to the spin-orbit interaction. For the complete energy solution this energy has to be combined with the relativistic mass correction shift, and one extra term (Darwin term) which applies only for $l = 0$ states. Both the spin-orbit interaction and the variation of mass with velocity depend explicitly on the quantum number l, but when combined they give an expression which depends only on the total angular momentum quantum number, j. Here the number j is obtained from the equations

$$\boldsymbol{j}^2\psi_T = j(j+1)\hbar^2\psi_T \tag{1.11}$$

where $\boldsymbol{j} = \boldsymbol{l}+\boldsymbol{s}$ is the total angular momentum and ψ_T is the total wave function including the spin dependence. When the Darwin term is included the total energy shift is given by the expression

$$\Delta E_{nj} = -\frac{\alpha^2 Z^2 E_n}{n^2}\left(\frac{3}{4}-\frac{n}{j+\frac{1}{2}}\right) \tag{1.12}$$

Figure 1.3 shows the energy level scheme of hydrogen ($Z = 1$) with the fine structure corrections included.

1.4. Hyperfine corrections and the Lamb shift

There are further corrections which can be made to the energy levels of single electron or hydrogen-like systems. First, the effects of the finite nuclear size and magnetic dipole moment can be considered. The former causes a shift in the energy levels whilst the latter causes a separation of the level into various components known as hyperfine structure. Both of these are much smaller than the fine structure and will not be considered in detail here. It is worth noting that hyperfine splitting, although small in energy, can significantly alter the relative populations of excited atomic states produced by the absorption of laser light.

Second, the level structure of hydrogen-like systems, even when calculated with the Dirac equation, does not completely agree with precise spectroscopic observations particularly for states with low n. The most famous discrepancy concerns the splitting of the $n = 2$, $j = \frac{1}{2}$ level of hydrogen. According to the Dirac equation the states with $l = 0$ (s state) and $l = 1$ (p state) should be

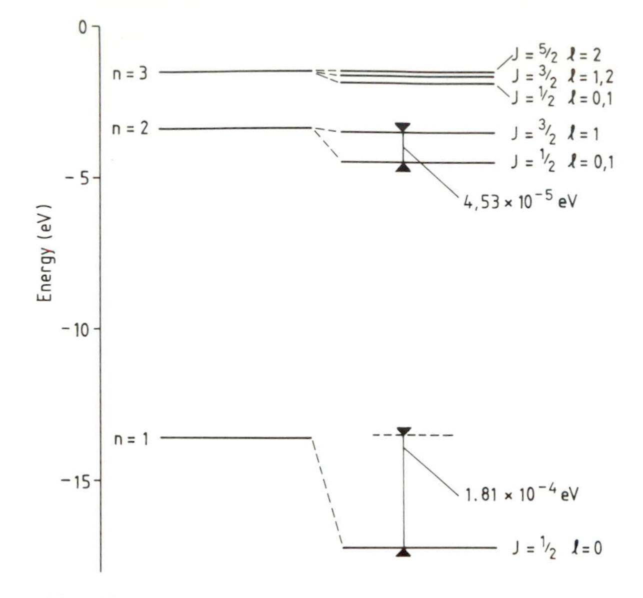

Figure 1.3. The atomic level structure of hydrogen calculated using Equation (1.5) and with the fine structure corrections (right). The fine structure shifts and splittings have been increased by a factor 2×10^4 in order to show them on the diagram.

degenerate but in a famous experiment carried out by Lamb and Rutherford (*Phys. Rev.* **72** (1947) p. 241) these two were found to be separated by $0{\cdot}035\,283\,\mathrm{cm}^{-1}$. This discrepancy was resolved only by recourse to quantum electrodynamics which will not be considered here.

1.5. Structure of many electron atoms

In an atom with many electrons each electron experiences both the attractive Coulomb force of the nucleus and the repulsive forces from all the other electrons. The Schrödinger equation can be written down in a similar way to Equation (1.2) but the potential would now include, as well as the central nuclear potential, a sum of terms from the Coulomb interaction of each electron with every other electron. Such an equation is impossible to solve analytically but can be solved numerically using techniques of successive approximations. The first step to describing the structure of the atom is known as the central field approximation. In this, it is assumed that each electron moves in a single central potential given by the equation

$$V(r) = \frac{-Z(r)e^2}{4\pi\varepsilon_0 r} \tag{1.13}$$

This is identical to the bare Coulomb potential of the nucleus except that there is now an effective charge $Z(r)$ which depends on the radial distance of the electron from the nucleus. When the electron is close to the nucleus the shielding of the other electrons is negligible so $Z(r)$ is equal to Z, whilst at very large distances the nucleus of charge Z combines with the $Z-1$ remaining electrons to make $Z(r)$ equal to unity. In between the form of $Z(r)$ is not known but it can be calculated using the Hartree method. This is an iterative procedure which starts by assuming some simple form for $Z(r)$, and then calculates the wave function for the many electron atom using the Schrödinger equation. This many-particle wave function, which of course describes the first approximation to the radial distribution of electrons in the atom, is then used to calculate a new form for $Z(r)$ which is in turn used to recalculate a new wave function. The procedure continues until the difference between several successive values of $Z(r)$, for every value of r, are small; that is until a reasonable convergence is established. The end result of this is to produce a wave function for the ground state of the atom which is formed from a set of electrons occupying single particle energy levels according to the exclusion principle. This principle demands that no two electrons can have the same wave function, or more explicitly, if the spatial wave functions are the same then the spin wave functions must be different. Since there are two possible values of the spin quantum number, and $2l+1$ different spatial wave functions corresponding to different values of m, each energy level of a given l can be occupied by $2(2l+1)$ electrons. The total wave function for the atom in the Hartree theory is then just given by the product

$$\psi_T = \pi\phi_a(1)\phi_b(2)\phi_c(3)\ldots \tag{1.14}$$

where $a, b, c, \ldots$ refer to the different single particle wave functions and (1), (2), (3), ... refer to the electrons. In actual fact the wave function shown in Equation (1.14), although formed in accordance with the exclusion principle (weaker condition), is incorrect since it is not properly antisymmetric. For fermions (spin $\frac{1}{2}$ particles like the electron) the total wave function must change sign but not magnitude under the exchange of any two particles, and it is easy to show that this is not true for the previous expression. If properly antisymmetric wave functions are used the procedure is known as the Hartree-Fock method. These calculations are more accurate since they include extra exchange forces which are not adequately taken care of in the simpler Hartree theory. They are of course also much more difficult and time consuming.

The set of single-particle energy levels generated by this procedure are similar to the energy levels of the single electron system except that they now depend not only upon the principle quantum number, n, but also on the orbital angular momentum number, l. This is not the same as the fine structure splitting but arises from the different average radial distributions of electrons in states of a given n but different l. Thus s states ($l = 0$), which are more concentrated near the nucleus, are lower in energy than p states ($l = 1$), where the electron spends more

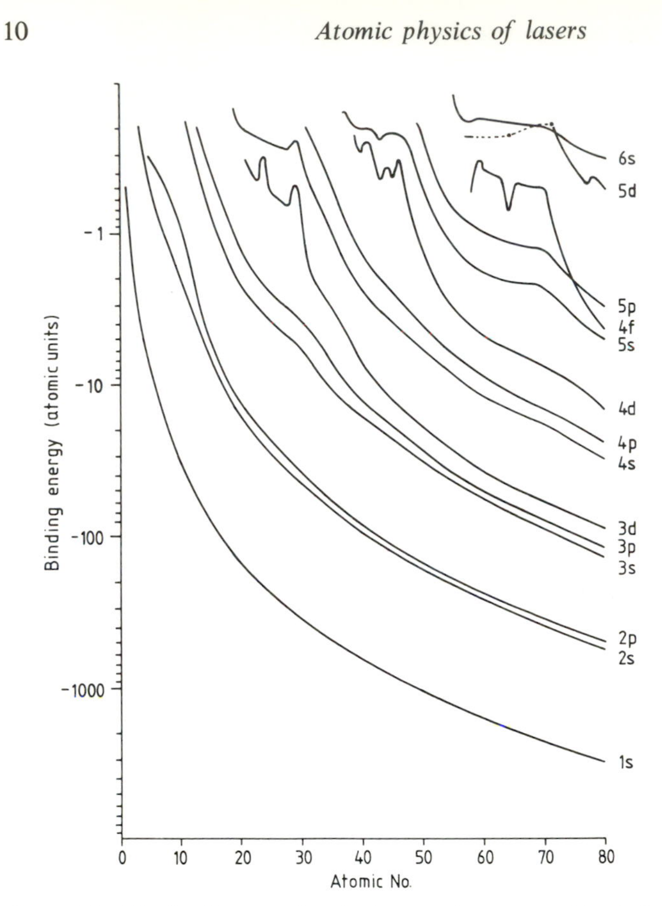

Figure 1.4. The binding energies of electrons in atomic orbits (n, l) obtained from numerical solutions of the Dirac-Fock equation. For clarity the fine structure splitting has been omitted, so each plotted point is the mean value of $j = l+\frac{1}{2}$ and $j = l-\frac{1}{2}$. (Values from J.P. Desclaux, Atomic Data and Nuclear Data Tables **12** (1973) p. 311.)

of its time further from the nucleus and is better screened from the nuclear charge by the presence of other electrons.

In Figure 1.4 the binding energy of electrons in different states is shown as a function of the atomic number. These are self-consistent calculations which use relativistic equations and are therefore different to those discussed previously, but they do illustrate the general ordering of levels. This level ordering can also be seen in the periodic table of the elements (Figure 1.5). Here the single particle levels are progressively 'filled' with electrons as the atomic number is increased. Chemical characteristics are determined by the number of electrons in the outermost or least bound single particle levels (or shells). Note that the single-particle energy spacing changes as a function of Z (Figure 1.4), and even the level ordering is altered in some cases. When the outermost unfilled single particle orbits are close together in energy the sequence of filling of levels can change. This occurs, for example, in the rare earth elements.

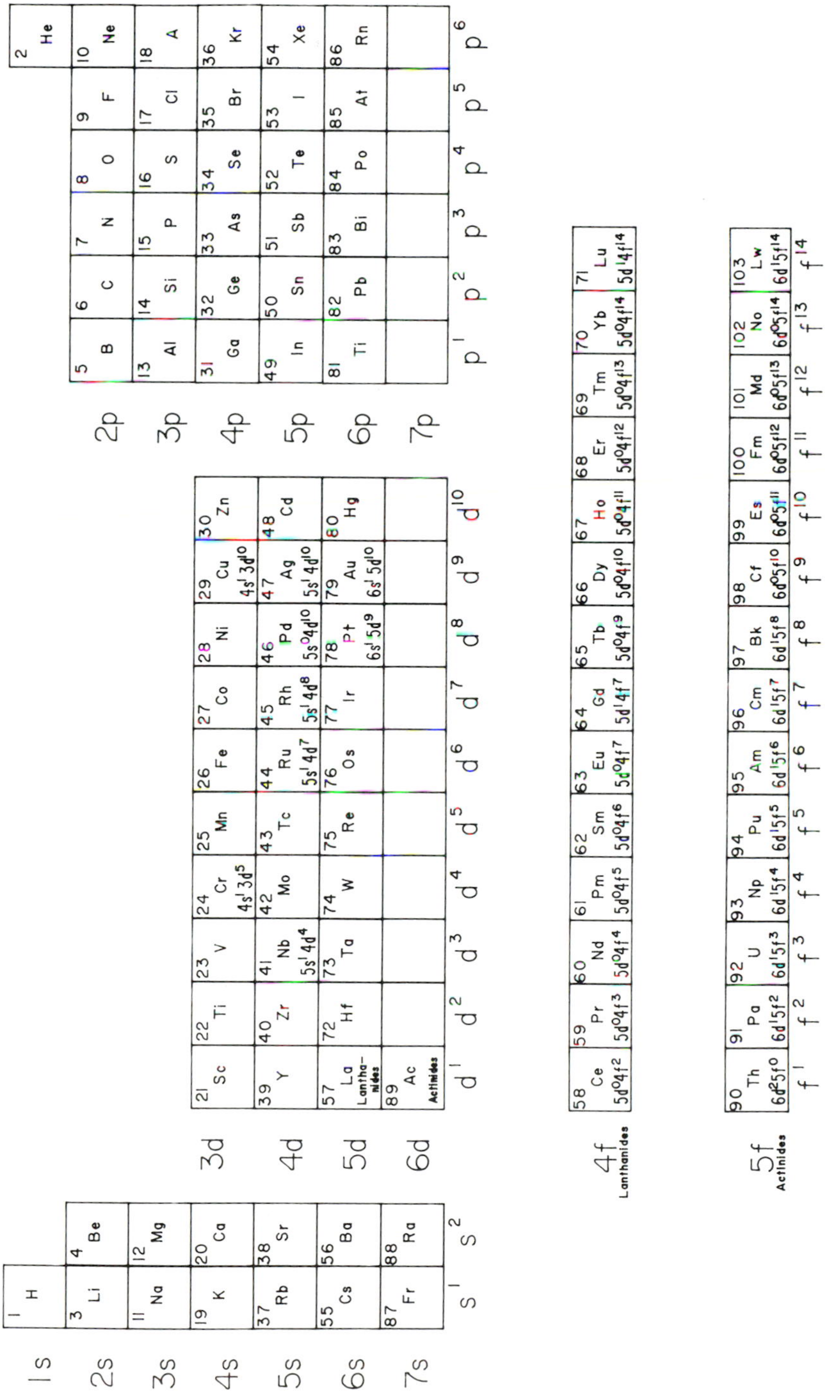

	s^1	s^2
1s	1 H	
2s	3 Li	4 Be
3s	11 Na	12 Mg
4s	19 K	20 Ca
5s	37 Rb	38 Sr
6s	55 Cs	56 Ba
7s	87 Fr	88 Ra

	d^1	d^2	d^3	d^4	d^5	d^6	d^7	d^8	d^9	d^{10}
3d	21 Sc	22 Ti	23 V	24 Cr $4s^1 3d^5$	25 Mn	26 Fe	27 Co	28 Ni	29 Cu $4s^1 3d^{10}$	30 Zn
4d	39 Y	40 Zr	41 Nb $5s^1 4d^4$	42 Mo	43 Tc	44 Ru $5s^1 4d^7$	45 Rh $5s^1 4d^8$	46 Pd $5s^0 4d^{10}$	47 Ag $5s^1 4d^{10}$	48 Cd
5d	57 La Lanthanides	72 Hf	73 Ta	74 W	75 Re	76 Os	77 Ir	78 Pt $6s^1 5d^9$	79 Au $6s^1 5d^{10}$	80 Hg
6d	89 Ac Actinides									

	p^1	p^2	p^3	p^4	p^5	p^6
1s						2 He
2p	5 B	6 C	7 N	8 O	9 F	10 Ne
3p	13 Al	14 Si	15 P	16 S	17 Cl	18 A
4p	31 Ga	32 Ge	33 As	34 Se	35 Br	36 Kr
5p	49 In	50 Sn	51 Sb	52 Te	53 I	54 Xe
6p	81 Ti	82 Pb	83 Bi	84 Po	85 At	86 Rn
7p						

	f^1	f^2	f^3	f^4	f^5	f^6	f^7	f^8	f^9	f^{10}	f^{11}	f^{12}	f^{13}	f^{14}
4f Lanthanides	58 Ce $5d^0 4f^2$	59 Pr $5d^0 4f^3$	60 Nd $5d^0 4f^4$	61 Pm $5d^0 4f^5$	62 Sm $5d^0 4f^6$	63 Eu $5d^0 4f^7$	64 Gd $5d^1 4f^7$	65 Tb $5d^0 4f^9$	66 Dy $5d^0 4f^{10}$	67 Ho $5d^0 4f^{11}$	68 Er $5d^0 4f^{12}$	69 Tm $5d^0 4f^{13}$	70 Yb $5d^0 4f^{14}$	71 Lu $5d^1 4f^{14}$
5f Actinides	90 Th $6d^2 5f^0$	91 Pa $6d^1 5f^2$	92 U $6d^1 5f^3$	93 Np $6d^1 5f^4$	94 Pu $6d^1 5f^5$	95 Am $6d^1 5f^6$	96 Cm $6d^1 5f^7$	97 Bk $6d^1 5f^8$	98 Cf $6d^0 5f^{10}$	99 Es $6d^0 5f^{11}$	100 Fm $6d^0 5f^{12}$	101 Md $6d^0 5f^{13}$	102 No $6d^0 5f^{14}$	103 Lw $6d^1 5f^{14}$

Figure 1.5. The periodic chart of the elements.

Some estimates can be made of the binding energies and sizes of the inner and outer orbits of electrons in atoms using the simple equations of the one electron case. For the 1*s* state there is only a small shielding of the atomic nucleus by the outer electrons. Here the approximate binding energy and radius can be found by substituting an effective nuclear charge of $Z-2$ in the one-electron equations. (It is left to the reader as an exercise in simple electrostatics to see why this should be approximately true.) Thus the binding energy is $(Z-2)E_H$ where E_H is the 1*s* electron binding in hydrogen (−13·6 eV), and the average radius is $a_0/(Z-2)$ where a_0 is the Bohr radius. For the least bound single particle level the effective charge is approximately equal to one. The binding energies and radii are roughly the same as the hydrogen atom single particle states with a value of n appropriate to the outermost unfilled level. Even though this crude estimate gives values for the outer radii which are too large, it still shows that atomic sizes do not, as is commonly misconceived, increase rapidly with atomic number. Note, however, that the inner orbits shrink much more rapidly as the atomic number is increased.

1.6. Residual interactions and excited atomic levels

The first approximation to the structure of the atom is obtained using only the single-particle wave functions generated from the central field. These wave functions and associated energy levels are designated by the quantum numbers n and l, and the ground state of the atom is given approximately by the configuration in which the electrons occupy the lowest levels according to the exclusion principle. The excited states in this simplified picture would be formed from configurations in which electrons are promoted from lower (or more tightly bound) states to higher ones. For a more accurate prediction of the ground state and excited states it is necessary to consider the effects of residual interactions. These are the interactions between electrons which cannot be incorporated into the central field.

It is not necessary to consider all the electrons (and energy levels) in these calculations since many of them occupy completely filled levels which are tightly bound. For certain atoms (He, Ne, Ar, Kr, Xe, ...) the highest single particle energy level is completely filled and there is a comparatively large energy gap to the next available level. These rare gas atoms define the boundaries of the major shells in atomic structure. Since single particle excitations are energetically unfavourable, their ground state structure is well described by the wave functions of the central field approximation, and the residual interactions essentially disappear. Also because these atoms are tightly bound, with all the electrons paired together, they are chemically inert and have a high ionization potential. In addition, they are characteristically small in size because the extra binding energy is gained at the expense of a shrinking of the outer shells.

Atoms with one electron 'outside' a closed major shell are conversely very chemically reactive. These are the alkali metals and their ground and excited states are well described by a single particle in the screened Coulomb potential arising from a central core of closed shells. For example, the ground states of Li, Na, K,... have a single electron in the 2*s*, 3*s*, 4*s*,... levels and the first excited state is produced by promoting this electron to the 2*p*, 3*p*, 4*p*,... levels respectively.

In the general case we need to consider several electrons (or holes for shells more than half full) outside the closed shell configuration. From these electrons and the available single-particle states a set of antisymmetric wave functions can be generated. If the interactions between the electrons are known it is possible to calculate, using perturbation theory, the excited states of the atom. For most atoms a suitable form for the interaction is given by the equation

$$\Delta H = \sum_{\substack{j,k \\ j \neq k}} \frac{e^2}{4\varepsilon_0 r_{jk}} + \frac{1}{\hbar^2} \sum_j \xi(r_j) \boldsymbol{l}_j \cdot \boldsymbol{s}_j \tag{1.15}$$

where the summation is over all the active electrons. The first term is just the Coulomb interaction between pairs of electrons whilst the second term is the spin-orbit interaction. Equation (1.15) refers only to the extra parts of the energy which are not included in the summation of the single particle energies. (The other two relativistic corrections can be added to the single particle energies.) It is therefore the extra term which needs to be added to the central potential.

1.7. LS coupling

The set of many-electron wave functions can be chosen in several ways, but the most useful, particularly for light atoms, is known as *LS*, or Russell-Saunders, coupling. In this scheme the individual spins, $\boldsymbol{s}_j$, and the orbit angular momenta, $\boldsymbol{l}_j$, are coupled separately together to form states with total orbital angular momentum $\boldsymbol{L}$ and total spin $\boldsymbol{S}$. $\boldsymbol{L}$ and $\boldsymbol{S}$ are thus given by the vector sums

$$\boldsymbol{L} = \sum \boldsymbol{l}_j \qquad \boldsymbol{S} = \sum \boldsymbol{s}_j \tag{1.16}$$

and the total angular momentum of the system $\boldsymbol{J}$ is then

$$\boldsymbol{J} = \boldsymbol{L} + \boldsymbol{S} \tag{1.17}$$

The total angular momentum, J, of any state must be well defined, and the wave function obeys a similar equation to (1.11) with J in place of j. Normally we say that J is a good quantum number of the state, and this must be so for any way of combining the single particle wave functions. Evidently the total angular momentum on the quantum axis, M_J, and the parity, P, are also good quantum numbers. For a pure central field interaction L and S are also good quantum numbers, that is they are well defined in any state. It can also be shown that, even when the individual Coulomb interactions between the electrons are added, L and S remain good quantum numbers. However, the second term in (1.15) mixes together different L, S states, but if this is much smaller than the Coulomb term L and S remain approximately good quantum numbers. This is the basis of the Russell-Saunders coupling, and, in these circumstances, the energy of the states can be calculated sequentially from wave functions of given L and S. Accordingly the energy changes due to the Coulomb interaction can be written as the diagonal matrix element

$$\Delta E_1 = \left\langle L, S, \alpha \left| \sum_{j \neq k} \frac{e^2}{4\varepsilon_0 r_{jk}} \right| L, S, \alpha \right\rangle = \int \Psi^*_{LS\alpha} \left(\sum_{j \neq k} \frac{e^2}{4\varepsilon_0 r_{jk}} \right) \Psi_{LS\alpha} \, dv \qquad (1.18)$$

where the integral is taken over all space. Ψ_{LS} is the total antisymmetric wave function of the configuration α. In the case of several electrons there will be more than one wave function of a given L, S that can be constructed from the configuration α. Wave functions are constructed from those of single electrons in the respective orbits $n_1 l_1, n_2 l_2, \ldots$. It is fairly easy to see that ΔE_1 will depend on the quantum numbers n and l, since these determine the spatial distributions of the electrons. It is more difficult to understand the dependence on the total spin, S. This arises because certain spin wave functions are related to the space wave functions by the requirement that the total wave function is always antisymmetric.

As an example of the above, consider a configuration of one electron in a $2p$ shell and one in a $3d$ shell. The total allowed values of the orbital angular momentum, L, are obtained from the vector addition of the two orbital angular momenta $l_1 = 1$ and $l_2 = 2$ according to the standard rules of quantum mechanics. That is, L has integral values from $|l_1 - l_2|$ to $|l_1 + l_2|$ which in this case gives $L = 1$, 2 or 3. States are labelled by the capital letters, S, P, D, F, $G, \ldots$ to denote L values of 0, 1, 2, 3, 4, ... In the same way S can have values from $|S_1 - S_2| = 0$ to $|S_1 + S_2| = 1$. In spectroscopic notation these are labelled by the multiplicity $2S+1$, and are known as singlets, doublets, triplets ... when $S = 0, 1, 2, \ldots$. This refers to the usual pattern of splitting when the LS interaction is included and is discussed later. For this example we have states (commonly called terms) ^1P, ^3P, ^1D, ^3D, ^1F and ^3F which arise from the single configuration $2p3d$. The splitting can be calculated from Equation (1.18) using properly antisymmetric wave functions constructed from $2p$ and $3d$ single particle wave functions.

When a particular configuration consists of more than one particle in the same orbit some of the values of L and S, which would be allowed by simple vector addition, are forbidden by the exclusion principle. The rules for this are fairly complicated when more than two electrons are involved and cannot be dealt with here. For two electrons in the same orbit a useful guideline is that $L+S$ must be even. So for example in the configuration np^2 (two electrons in a np shell), the possible terms are ^1S, ^3P and ^1D, and the ^3S, ^1P and ^3D are not allowed.

For configurations involving three or more electrons there can be several terms with the same values of L and S. These will have different total wave functions and hence different energies. In these cases it is helpful to show how the coupling has been made. The general configuration and term for three particles is then written $nln'(^{2S+1}L)n''L''\,^{2S'+1}L'$ signifying that the electrons in orbits l and l' are first coupled to form the term ^{2S+1}L, which is then coupled to produce the final term $^{2S'+1}L'$. The bracketed term is known as the parent. For example a $npn'dn''p$ configuration can give rise to terms ^2G, ^2F and ^2D from the parent $npn'd(^1F)$, and terms ^2P, ^2D and ^2F from the parent $npn'd(^1D)$. In fact there are four possible ^2F terms and each will have a different energy.

After calculating the gross level scheme from the single particle state (configurations), and then the splitting, arising from the residual Coulomb interaction, the effect of the spin-orbit interaction can now be considered. In the LS approximation, this causes a fine structure splitting of each of the previous terms. Even though the perturbation given by the second part of Equation (1.15) can mix terms with different L and S, this mixing is small because the states are comparatively widely separated in energy. (In perturbation theory the second-order terms which determine the mixing of states have this energy separation in their denominator.)

The second part of Equation (1.15) is now written as a single expression $\xi(\alpha, L, S)\,\boldsymbol{L}.\boldsymbol{S}$, where the constant depends on the particular configuration and the term within that configuration. This constant can be evaluated using the complete wave functions. Since $\boldsymbol{J} = \boldsymbol{L}+\boldsymbol{S}$ then $\boldsymbol{J}^2 = \boldsymbol{L}^2+2\boldsymbol{L}.\boldsymbol{S}+\boldsymbol{S}^2$ and

$$\Delta H_2 = (1/2\hbar^2)\xi(\alpha, L, S)(\boldsymbol{J}^2-\boldsymbol{L}^2-\boldsymbol{S}^2)$$

then

$$\Delta E_2 = (1/2)(\xi(\alpha, LS)(J(J+1)-L(L+1)-S(S+1)) \tag{1.19}$$

States of a given L and S are split according to J. The interaction therefore removes the J degeneracy. For singlet states, $S = 0$, the interaction vanishes and there is no splitting. This corresponds to the case when all the components in the sum, $\sum \xi(r_j)\boldsymbol{l}_j.\boldsymbol{s}_j$ can be paired off exactly. Thus each component $l_j s_j$ has an equal and opposite one $-l_j s_j$, corresponding to the different alignment of the vectors. For doublet terms, J is either $|L-\frac{1}{2}|$ or $|L+\frac{1}{2}|$ so there are normally two levels except when $L = 0$ (s states) where again the interaction vanishes. Similarly triplet terms have three components with $J = |L-1|$, L and $|L+1|$ except when $L = 0$ when there is only one level with $J = L$. Each term is therefore split into $2S+1$ (or $2L+1$ if L is less than S) fine structure components or levels. The separation of these levels can be obtained from Equation (1.19), and is given by the expression

$$E(J)-E(J-1) = \xi J \tag{1.20}$$

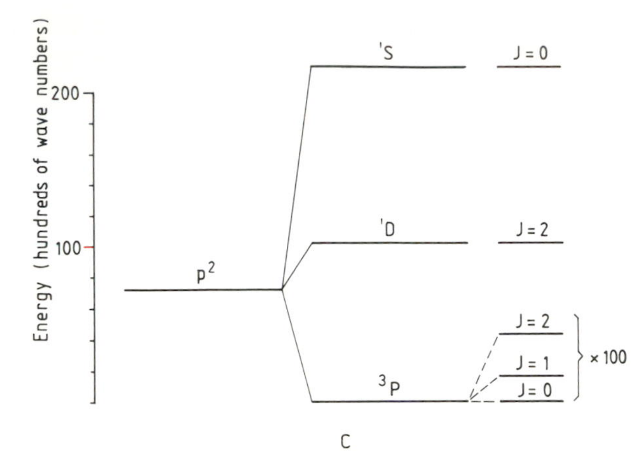

Figure 1.6. Energies of the levels in the carbon atom arising from the configuration $2p^2$. The left hand column shows the unperturbed configuration which is split by the residual Coulomb interaction into three terms, ^3P, ^1D and ^1S. The triplet, ^3P, is further separated by the spin-orbit interaction.
(Note that the Landé interval rule is not quite followed.)

This is known as the Landé interval rule. It can be used in spectroscopy to establish the validity of Russell-Sanders coupling, or if the interval is established, identify the values of J (Catch 22?).

As an example of *LS* coupling, Figure 1.6 shows the energy levels of carbon which arise from the configuration $2p^2$. Note that the interval rule for the ground state splitting is not obeyed exactly.

1.8. Configuration mixing

In describing the ground and excited states of atoms many configurations often need to be considered. The simplest set of configurations occur for the alkali metals where the levels can be described by one electron in different single particle states. So that for Li the ground state would be described by an electron in the $2s$ shell, and the excited states in order of increasing energy are formed when this electron is promoted to the $2p$, $3s$, $3p$, $3d$, $4s$, $4p$ . . . levels. In Na a similar sequence occurs, except that the ground state is now $3s$ and the excited states are $3p$, $4s$, $3d$, $4p$, ... Note that in sodium the level spacing and even the ordering alters because the central potential is not the same. The $3d$ level is now above the $4s$ level because it is more effectively screened from the nucleus by the other electrons. (In potassium the $5s$ state is below the $3d$ state!)

Atoms further away from closed shells have a higher density of excited atomic levels and their description requires more configurations. For example, the excited states of the carbon atom require 25 configurations including the obvious ones of $2p^2$, $2p3s$, $2p3p$, $2p4s$, $2p5s$, $2p3p$, $2p4p$, $2p3d$ and $2p4d$. The number of configurations in atoms with more valence electrons can be quite staggering.

In these circumstances it is sometimes not possible to describe an excited state (or even a ground state) by a single configuration. This mixing is brought about by the residual interactions and is larger when the unperturbed configurations are close together. Only configurations of the same parity can be mixed. Thus in the previous example an even-parity level might contain only the configurations $2p^2$, $2p3p$, $2p4p$, ... and an odd-parity level the configurations $2p3s$, $2p4s$, $2p3d$, ...

1.9. *j–j coupling*

Although the *LS* coupling scheme is the most useful way to categorize atomic energy levels it is not applicable when the spin-orbit force becomes larger than the residual electrostatic interaction. This situation occurs for certain electron configurations in heavy atoms, and here the most useful categorization of the states is by *j–j* coupling. Because the spin-orbit force is much larger the electrons appear to move quite independently of one another, and in these circumstances the individual values j_i, l_i and s_i are approximately good quantum numbers. States are labelled by these numbers as well as the total angular momentum derived from the vector sum

$$\boldsymbol{J} = \sum_i \boldsymbol{j}_i \tag{1.21}$$

The normal notation for a state is $(l_{1j_1}, l_{2j_2}, l_{3j_3}, \ldots)_J$. For example, a $5p6s$ configuration has four levels $(p_{1/2}, s_{1/2})_0$, $(p_{3/2}, s_{1/2})_1$, $(p_{3/2}, s_{1/2})_1$ and $(p_{3/2}, s_{1/2})_2$. The largest splitting is between the two levels with $p_{1/2}$ components and $p_{3/2}$ components, and is due to the spin-orbit interaction. Final separation according to J is brought about by the residual electrostatic interaction. Clearly the cause of fine structure has been reversed from the *LS* case!

When the two perturbations, spin-orbit and electrostatic, are comparable neither of these schemes is applicable. Here only J, M and parity are good quantum numbers and the calculations are more complicated involving secular equations.

1.10. *Molecular structure*

Since molecules possess many more degrees of freedom their structure is much more complicated than atoms. In this section a brief account will be given of the basic physics of *diatomic* molecules. The extension to larger molecules is one of complexity rather than conception. It will concentrate on the structure of excited states since this is the most relevant aspect for laser physics.

1.11. Chemical bonding

The two atoms in a molecule may be bound together in an ionic or a covalent bond, or in some intermediate way. In an ionic bond an electron from one atom is transferred to the outer orbit of the other atom, thus making a positive and negative ion which are bound by electrostatic forces. A common example is NaCl, where the loosely bound 3*s* electron from sodium is transferred to the 3*p* vacancy (hole) in chlorine. The energy deficiency, that is the difference between the binding energy of the electron in sodium and the electron affinity of chlorine is made up by the extra electrostatic energy. In terms of the electronic structure the molecule resembles the two inert gases argon and neon joined together, with the proviso that the outer shells will have different sizes in accordance with the altered nuclear charge.

In a covalent bond the valence electrons are effectively shared between the two atoms, that is the outer electrons, instead of moving in the central potential of their separate atoms, now move in a two-centred potential of the combined system. A simple example is the H_2 molecule where the two electrons occupy a single orbital (single-particle energy level of a molecule) of two separated protons. For hydrogen atoms the electrons are in the 1*s* state and are bound by 13·6 eV, but in the composite molecule each electron becomes more bound because it comes under the influence of two protons. This extra binding exceeds the energy required to overcome the Coulomb repulsion of the protons so the total system is bound, and the variation of the energy as a function of internuclear separation shows a distinct minimum. At smaller distances the Coulomb repulsion dominates and the total energy rapidly becomes positive, whilst at larger distances it tends to the sum of the separate electron binding energies.

The electron wave function describing the H_2 molecule is spatially symmetric about the plane bisecting the internuclear axis, and is formed from two electrons in the lowest orbital of the molecule. In these circumstances the electron spins are antiparallel and the state is a singlet. For this configuration the electron density is high at the mid point between the two nuclei, and this serves to increase the individual Coulomb binding energies of the electrons whilst reducing the opposing effects of nuclear repulsion. It is thus more energetically favourable than the nearby triplet state which is not bound.

In larger molecules the binding arises in a similar way except that only the valence electrons are shared, while the remaining, more tightly bound electrons, still belong to one atom, though there will be Stark splitting of the levels. An important point, which is discussed more thoroughly below, concerns the nature of the single particle states in the molecular system. The total number of states in the combined molecular system is equal to the number of states in the separated atoms (Ehrenfest adiabatic law). In general, the lowest states of the molecule can be associated with the lowest single particle states of the separated atom. So, for example, a sodium atom has a single 3*s* valence electron which can couple with an electron from another atom to form a molecular singlet-state bond with two electrons in the lowest molecular orbital. An atom like barium has two electrons in the 3*p* single particle level and can form two singlet-state bonds (homopolar bonds). Clearly, the chemical bonding can be related in a qualitative way to the shell structure of atoms.

1.12. Single particle molecular states

Consider the case of a single electron moving under the influence of two bare nuclei with Z_1 and Z_2 protons. When the separation of the two nuclei is zero the energy levels will be those of the single electron system with $Z_1 + Z_2$ protons. At small separations there will be, in addition to the spherically symmetric central potential, an extra term which has the same effect as adding a unidirectional electric field to the atom. In these circumstances the splitting of the levels will be similar to the Stark splitting in an external field. Each single particle level with quantum number n is split into $l+1$ states according to the value of the new quantum number λ where $\lambda = |m_l|$. λ has values 0, 1, 2, . . . up to a maximum value l, and these are designated by the Greek letters σ, π, δ, ... The states are labelled according to their origin and the value λ, and so the general sequence for a single electron is $1s\sigma$, $2s\sigma$, $2p\sigma$, $2p\pi$, $3s\sigma$, $3p\sigma$, $3p\pi$, $3d\sigma$, $3d\pi$, $3d\delta$, $4s\sigma$, ... At large separations, the energy levels are the same as the two separated atoms, but also with the addition of a Stark shift caused by the proximity of the other atom. In between the levels still retain the quantum number λ (l and even n are no longer good quantum numbers) and each will be associated with a particular state of the two limits. Mathematically this amounts to saying that the solutions must give the limiting values, and we can draw an energy diagram (Figure 1.7). In joining the lines, the rule is that a σ orbital on the left must go to a σ orbital on the right and a π orbital on the left must go over only to a π orbital and so on. Also, as long as λ is the only good quantum number at intermediate distances (which is true for almost all cases except H_2^+), the lowest orbitals on the left of a given λ must join with the lowest orbitals on the right of a given λ. When the nuclei have equal charges an extra symmetry about the plane bisecting the atoms is introduced and the joining is somewhat different especially for the lower levels.

Diagrams like Figure 1.7 will still be applicable for the outer electrons of a composite atomic system, except that the single particle spacings will be altered by the screening of the nuclei from the tightly bound electrons.

To obtain the total binding energy of the atom we need to add the Coulomb repulsion of the nuclei to the electron energies. This is always positive and increases as the internuclear separation is reduced. Molecules can be bound only

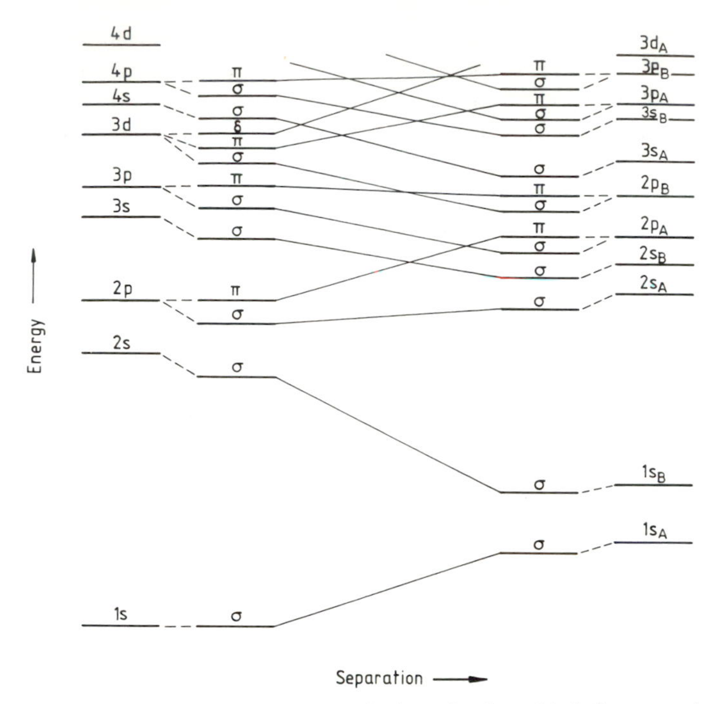

Figure 1.7. Schematic diagram of the energies in molecular orbitals for unequal nuclear charges. The scale on the separation axis is non-linear.

if the electrons in the orbitals of the atoms decrease in energy as the atoms come close together. If two electrons are in the lowest orbital of the system shown in Figure 1.7 there is a decrease in energy corresponding to the slope of the line going from right to left and the system will be bound. Adding a third electron would reduce this energy since the next available level slopes in the opposite direction. (Each of the levels is two-fold degenerate except for σ states where there is only one spatial wave function). In this way electrons are called bonding or antibonding according to the slope of the line, and roughly speaking the molecule will be stable if the number of bonding electrons exceeds the number of antibonding electrons. For H_2 the second single particle level slopes upwards in going from right to left (similar to Figure 1.7) and this causes the triplet state, formed from a single electron in each of the lowest orbitals, to be unbound.

1.13. Coupling of the electrons in single particle orbitals

The coupling of the electrons in the single particle orbitals will now be considered. As with atoms the gross structure of the excited states can be evaluated by adding the single particle energies. In molecules, however, this has to be done at all separations of the atoms because the minimum energies will not occur at the same separation for each configuration. This implies that the potential energy as a function of internuclear separation is different for each configuration. In fact, it will be different for each term within the configuration.

The orbital angular momenta $l_1, l_2, l_3, \ldots$ of the individual electrons are no longer good quantum numbers. Instead the states are labelled $\lambda_1, \lambda_2, \lambda_3, \ldots$ and, from these, states can be formed with good orbital angular momentum *about the internuclear axis* using the sum

$$\mathbf{\Lambda} = \sum_i \boldsymbol{\lambda}_i \tag{1.22}$$

where the summation extends over all the orbital electrons. Since the vectors $\boldsymbol{\lambda}$ all lie in the symmetry axis the values are obtained by simple algaebraic addition, noting that either sign is possible. This corresponds to the two values of m_l. According to convention, the numbers Λ are labelled by the Greek letters $\Sigma, \Pi, \Delta, \Phi, \ldots$ corresponding to the values 0, 1, 2, 3, ... Λ is analogous to L in the atomic case, but there are in general fewer values for a given set of individual quantum numbers. So for two unequal values λ_1 and λ_2, Λ can be either $|\lambda_1 - \lambda_2|$ or $|\lambda_1 + \lambda_2|$ and each is two-fold degenerate corresponding to the different directions of Λ. Of course, when either λ_1 or λ_2 is zero, there is only one value of Λ. For $\lambda_1 = \lambda_2$ ($\lambda_1 \neq 0$), $\Lambda = 0$ but there is still a two-fold degeneracy. When both are zero there is only one spatial wave function.

This two-fold degeneracy remains even when the residual interactions are included *except* for Σ states ($\Lambda = 0$) where the states are labelled Σ^+ and Σ^-. These refer to the symmetry of the wave function when reflected in a plane through the line joining the nuclei.

In the case of a weak spin-orbit interaction the total spin, S, is still, approximately, a good quantum number, and we can couple together the spins separately to form singlets, doublets, triplets, etc. (See Equation (1.16).) For example, consider two π electrons from *non equivalent* orbits. These can form states $^1\Sigma^+$, $^3\Sigma^+$, $^1\Sigma^-$, $^3\Sigma^-$, $^1\Delta$ and $^3\Delta$ and these will be split by residual electrostatic interaction, given by the first term of Equation (1.15). When the electrons are in the same orbital there are restrictions, similar to those encountered in atoms, arising from the Pauli principle. Thus, if the λ's are parallel the spins must be antiparallel, so for two equivalent π electrons there is no $^3\Delta$ state. Additionally, the symmetry properties require that there will only be two $\Lambda = 0$ terms, $^1\Sigma^+$ and $^3\Sigma^-$. Note that when there are four equivalent π electrons the subshell is closed and there is only one term, $^1\Sigma^+$. Lists of terms for more complicated cases can be found in detailed texts.

When the two atoms are identical there is an extra symmetry about the plane bisecting the two nuclei. A similar diagram to Figure 1.7 can be drawn except that the two levels, which might be expected to overlap, are separated according to whether they are antisymmetric (g) or symmetric (u). In combining several electrons the terms of a given configuration are all even or odd according to the number of electrons.

The term constructed from the electrons in the orbitals are further split by the spin-orbit interaction in a similar way to atoms. These are labelled according to the total value of the angular momentum about the internuclear axis, Ω. Ω is given by the equation

$$\Omega = |\Lambda + \Sigma| \tag{1.23}$$

where Σ is the total spin angular momentum about the nuclear axis. Σ has integral values from $-S$ to $+S$, and in contrast to Λ, positive and negative values are relevant. (The spin-orbit interaction depends on the relative directions.) To a first approximation the energy of a multiplet term is

$$E = E_0 + A\Lambda\Sigma \tag{1.24}$$

where E_0 is the value when the electron spin is neglected, and A is a constant for a given term. There are $2S+1$ components except when $\Lambda = 0$ and, unlike the atomic case, this is true even when S is larger than Λ. An important point to realize is that the total angular momentum, J, is the same as Ω.

For large spin interaction an alternative coupling scheme, similar to *j–j* coupling in atoms, must be used. Details can be found in more advanced texts.

1.14. Vibrational and rotational excited states

Molecules, unlike atoms, do not have spherical symmetry and rotations about any axis perpendicular to the line joining the two nuclei give rise to excited energy levels. In addition, the atoms which form the molecule are not held rigidly together, but can vibrate in antiphase with respect to their common centre of gravity.

We consider first the vibrational energy levels of a diatomic molecule. When the restoring force on each atom is proportional to the atomic separation from the equilibrium position, the Schrödinger equation is written

$$\frac{d^2\psi}{dx^2} + \frac{8\pi^2}{h^2}\mu(E - \tfrac{1}{2}kx^2)\psi = 0 \tag{1.25}$$

where $\mu = m_1m_2/(m_1+m_2)$ is the reduced mass, x is the difference between the separation and the equilibrium separation and k is the force constant. This is the standard quantum mechanical harmonic oscillator equation and the energies of the states are given by the equations

$$E_v = (n + \tfrac{1}{2})h\nu_0 \tag{1.26}$$

and

$$\nu_0 = \frac{1}{2\pi}\sqrt{\frac{k}{\mu}}$$

where ν_0 is the classical vibration frequency. The vibrational quantum number, n, can take only integral values 0, 1, 2, 3, ... In a typical molecule like HCl the force constant is approximately 5×10^5 dynes/cm (500 N/m) so the separation between levels, $h\nu_0$, is $2{\cdot}9 \times 10^3\ \text{cm}^{-1}$ or 0·35 eV. This is much less than the first electronic excited state ($^1\Pi$) at $7{\cdot}7 \times 10^4\ \text{cm}^{-1}$. When the vibrational quantum number is zero the molecule still has an energy $\frac{1}{2}h\nu_0$, which is called the zero-point energy. It is a pure quantum mechanical effect and has no classical counterpart. Interestingly, the existence of a zero-point energy can be verified directly by measuring the difference between the dissociation energy of molecules with unequal isotopic composition. For example D_2, DH and H_2 have different dissociation energies because μ is altered whilst the restoring force constant, which depends only on the chemical identity of the molecule, is fixed.

It is also found, empirically, that the spacing of vibrational levels is not constant as Equation (1.26) predicts. The level spacings decrease with increasing n, and this is associated with changes in the potential energy curve from a true harmonic (x^2) form. These effects can be calculated by introducing a more realistic form for the potential energy into Equation (1.25).

Consider now the energy levels generated by a rotating diatomic molecule. Only rotations normal to the symmetry axis have any meaning, so the energy for a 'rigid' rotator can be written

$$E = \boldsymbol{R}^2/2I \tag{1.27}$$

where I is the moment of inertia and $\boldsymbol{R}$ is the rotational angular momentum. Now since the angular momentum is quantized, $\boldsymbol{R}^2$ is equal to $R(R+1)\hbar^2$ and the energy levels associated with the rotation are given by the equation

$$E_R = \frac{R(R+1)\hbar^2}{2I} \tag{1.28}$$

with $R = 0, 1, 2, \ldots$. The spacing between adjacent levels, $E_R - E_{R-1}$, is $R\hbar^2/I$. For the molecule HCl, the moment of inertia can be calculated from the interatomic spacing and the first excited rotational level ($J = R = 1$) is $20{\cdot}8\ \text{cm}^{-1}$ ($2{\cdot}6 \times 10^{-3}$ eV) above the ($J = R = 0$) ground state. This is small in comparison with the average thermal energies of the molecule at room temperature. According to the Maxwell–Boltzmann distribution law, all the states up to $J = 10$ will be significantly populated in this case. The vibrational spacing for HCl is 140 times larger so the first excited vibrational level is not populated by thermal collisions at room temperature.

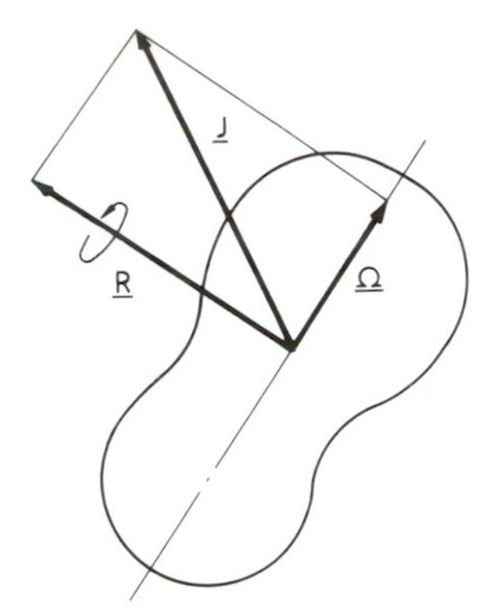

Figure 1.8. Vector coupling of the angular momentum about the internuclear axis, **Ω** (due to the electronic motion and spin), with the rotational angular momentum, ***R***, to form a state with total angular momentum, ***J***.

The wave functions associated with the rotational motion are given by normalized surface harmonics. These are identical to the angular solutions of the hydrogen atom, except that l and m are replaced by R and its component, M_R. Such functions have even or odd parity according to the value of R (l in the hydrogen case), so the total angular momenta, J, and parities of a rotational band built on a $J = 0$ ground state are $0^+, 1^-, 2^+, 3^-, \ldots$. For a symmetrical molecule ($N_2, O_2, \ldots$) the negative parity rotational wave functions are prohibited so that only the even numbers of this band are generated.

In deriving Equation (1.28), the coupling between the rotational and the electron motion has been neglected. When the electronic state has a non zero total angular momentum this cannot be ignored. The simplest case to consider is known as Hund's case (a), and Figure 1.8 shows a vector diagram for the coupling of the different components of angular momentum. It is assumed that, even in the case of a rotating system, the electron angular momentum about the internuclear axis, Ω, is well defined. In this case the rotational energies are given by the equation

$$E_{RJ} = \frac{\hbar^2}{2I} J(J+1) + \left(\frac{\hbar^2}{2I_e} - \frac{\hbar^2}{2I}\right)\Omega^2 \tag{1.29}$$

$\Omega = |\Lambda + \Sigma|$ is the resultant electronic angular momentum about the symmetry axis and I_e is the associated moment of inertia. Evidently I_e will be much smaller than I. The rotational band generated by this equation is the same as before except now the lowest energy state is when J equals Ω, and all the states are shifted by a constant amount proportional to Ω^2. Equation (1.29) is equivalent to (1.28) when Ω is zero.

As an example, consider the multiplet of electronic levels $^3\Delta_1$, $^3\Delta_2$ and $^3\Delta_3$. On each level will be built a rotational band with the lowest members having J

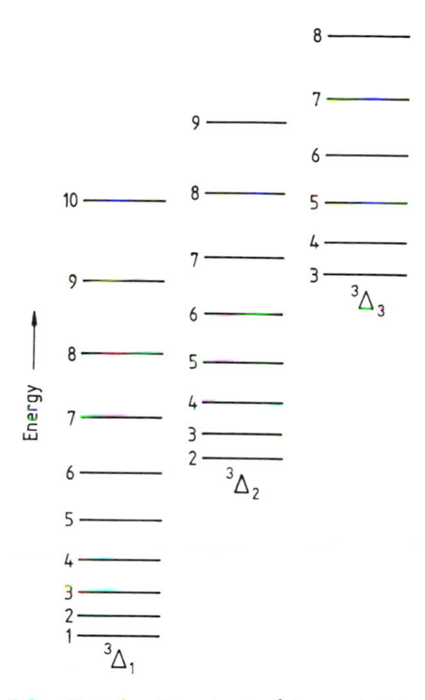

Figure 1.9. Rotational levels of a $^3\Delta$ term in Hund's case (a). The diagram has been constructed using the correct interval spacing as determined by Equation (1.24).

equal to 1, 2 and 3 respectively and spacings determined by Equation (1.29). The level structure is shown in Figure 1.9. More complicated cases of coupling can be found elsewhere.

So far the discussion has assumed that the molecule is rigid and remains so even when rotating rapidly. In reality, centrifugal stretching causes the moment of inertia to increase as J is increased. The energy spacing between adjacent levels decreases from that predicted by the previous equation. This effect is quite small and can be accounted for by the addition of a term proportional to $J^2(J+1)^2$. This centrifugal stretching can also be predicted from the form of the potential energy as a function of nuclear separation, which in turn can be obtained from the spacing of the vibrational energy levels.

1.15. Combining the electronic, vibrational and rotational energies

Any molecular state can be made up of electronic, vibrational and rotational excitations and it is necessary to consider how these are combined. In fact, a very good approximation is to consider that the total wave function can be made up of (antisymmetric) products of the three separate wave functions describing each type of motion. This amounts to saying that the total energy is given by

$$E_T = E_e + E_v + E_{RJ} \tag{1.30}$$

where E_e, E_v and E_{RJ} are the electronic, vibrational and rotational energies found from the previous analysis. Mathematically, such a solution results from ignoring certain cross-coupling terms in the Hamiltonian of the Schrödinger equation. Physically we can imagine that the coupling between the electronic and vibrational/rotational states is small because electrons, whose velocities are large, can adjust rapidly to the changing potential in which they move. Similarly, coupling between rotation and vibration can be ignored because the vibrational frequencies are much higher than the rotational ones. This is reflected in the much larger spacing of vibrational levels. Figure 1.10 shows the excited states of an idealized molecule where no coupling exists between the different motions. Of

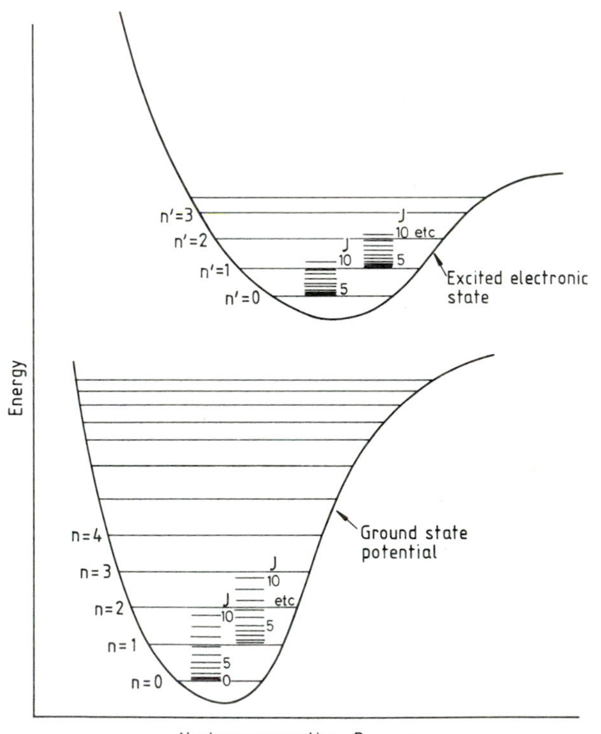

Figure 1.10. A schematic energy level diagram for a molecule with no coupling between the electronic, vibrational and rotational degrees of freedom. For simplicity, both of the electronic terms are $^1\Sigma^+$. Note that the intercept of the vibrational lines with the potential energy curve gives the classical turning points of the vibrational motion which do not, for a quantum mechanical harmonic oscillator, rigidly define the extremities of the motion.

course, in reality some coupling does occur, particularly when the electronic levels are close together. An example of this was mentioned in the last section.

Although the interaction of electromagnetic radiation with atoms and molecules is properly dealt with in Chapter 2 it is worthwhile mentioning two important points concerning the decay of excited molecular states. Firstly, the total parity of the molecule must change when it emits radiation (excluding forbidden transitions). This means that pure vibrational states built on a single electronic configuration cannot decay to each other. However the rotational levels built on each vibrational state have alternate parity assignments, and so decay is possible from certain members of one rotational band to another. Secondly, allowed transitions between rotation-vibration levels, based on the same electronic structure, are possible only when the molecule has a permanent dipole moment. Symmetric molecules like O_2, N_2 etc. have no dipole moment and this significantly alters their emission/absorption spectra.

Bibliography

Condon, E.U. and Shortley, G.H., 1963, *The Theory of Atomic Spectra* (Cambridge University Press).

Heitler, W., 1956, *Elementary Wave Mechanics* (2nd edition) (Oxford University Press).

Herzberg, G., 1950, *Spectra of Diatomic Molecules* (2nd edition) (New York: Van Nostrand).

Pauling, L. and Wilson, E.B., 1935, *Introduction to Quantum Mechanics* (New York: McGraw-Hill).

Schiff, L.I., 1948, *Quantum Mechanics* (New York: McGraw-Hill).

Willmott, J.C., 1981, *Atomic Physics* (Chichester: Wiley).

Woodgate, G.K., 1980, *Elementary Atomic Structure* (2nd edition) (Oxford University Press).

CHAPTER 2

the interaction of electromagnetic radiation with atoms

2.0. Introduction

This chapter deals with the interaction of electromagnetic radiation and atoms using a semiclassical description, although some references will be made to the more powerful techniques of quantum electrodynamics. It is mainly concerned with the simplified case of a single excited atomic state which can be populated by the absorption of light and subsequently returns to the ground state by stimulated and spontaneous emission. (Three and four level systems, which can give rise to light amplification, are considered in Chapter 3.) Most of the analysis is for isolated atoms although some references are made to line broadening effects in gases in Section 2.9.

The chapter is roughly divided into four sections. Section 1 considers the derivation of the Planck black body radiation equation. Section 2 derives the Einstein coefficients for absorption by considering the radiation field as a perturbation to the quantized atomic states. Section 3 considers the coupled equations for absorption and emission and relates these to the phenomenological Einstein equations, whilst Section 4 relates the structure of atomic states to their electromagnetic decay properties.

2.1. Planck's law and the quantization of the radiation field

In this section the spectral distribution of radiation from a black body will be derived by quantization of the electromagnetic field. A black body is a system at a uniform temperature which absorbs all the radiation incident upon it. Experimentally this can be realized using a well insulated oven with a small hole in it. Radiation emitted from the hole can be measured, whilst all the energy incident upon it is totally absorbed. For this idealized situation, the spectral energy density within the cavity and the outward flow of energy† per unit area of the hole are dependent only on the temperature of the oven. In addition, the spectral distribution of radiation exhibits a maximum at a particular wavelength, and this maximum shifts to shorter wavelengths as the temperature is raised. The

† In practical situations, the emissivity of the surface has to be considered.

failure of the classical method to predict the shape of this spectrum marked a turning point in the history of modern physics, and led to Planck's hypothesis of the quantum. Planck originally derived a correct form for this spectrum by considering the interaction of a classical electromagnetic field with quantum 'oscillators' (representing the atoms) of discrete frequencies ε_0, $2\varepsilon_0$, $3\varepsilon_0, \ldots$ where $\varepsilon_0 = h\nu$, and ν is the frequency of the radiation. However, a more satisfactory explanation, which produces the same result, is obtained by direct quantization of the electromagnetic field, and this approach will be used here.

Firstly, we need to consider, classically, the modes of the electromagnetic field. With each mode we can associate a certain energy and hence quantize the field. To derive the modes, a box with perfectly reflecting walls is introduced into the cavity. This will not affect the energy distribution because there can be no net flow of energy across any area within the cavity, if the temperature is uniform. The radiation will then be trapped within the box in the form of standing waves, and it is the distribution of these modes in frequency space which are required for the calculation. Inside the cavity box the magnetic and electric fields of the radiation must be solutions of Maxwell's equations (in free space) with the appropriate boundary conditions on the walls. In these circumstances the electric field components are given by the equations

$$\begin{aligned} E_x &= E_x(t)\cos(k_x x)\sin(k_y y)\sin(k_y z) \\ E_y &= E_y(t)\sin(k_x x)\cos(k_y y)\sin(k_z z) \\ E_z &= E_z(t)\sin(k_x x)\sin(k_y y)\cos(k_z z) \end{aligned} \tag{2.1}$$

where the wavevector, $\boldsymbol{k}$, is related to the length of the cube, L, by the relations

$$k_x = n_x \pi/L \qquad k_y = n_y \pi/L \quad \text{and} \quad k_z = n_z \pi/L \tag{2.2}$$

and n_x, n_y and n_z can have integral values 0, 1, 2, 3,.... Evidently, Equation (2.1) gives the correct boundary conditions in that the electric field vanishes on all the faces of the cube. Also, for each set of values n_x, n_y and n_z, there are two modes corresponding to the two opposite directions of $\boldsymbol{E}(t)$. (An alternative solution to Maxwell's equations is (2.1) with all the signs reversed.)

Figure 2.1 shows a plot of the allowed values of the wavevector, $\boldsymbol{k}$. To find the number of modes per unit interval of k space, $U(k)$, a shell of radius k, and thickness δk, is drawn on the figure. The number of modes in the small interval δk, is then twice the number of points n_x, n_y and n_z, which lie within this shell. Since a single point occupies a volume $(\pi/L)^3$, $U(k)$ is given by the expression

$$U(k)\,\delta k = 2 \times \frac{1}{8}\frac{(4\pi k^2)\,\delta k}{(\pi/L)^3} = (k^2/\pi^2)L^3\,\delta k \tag{2.3}$$

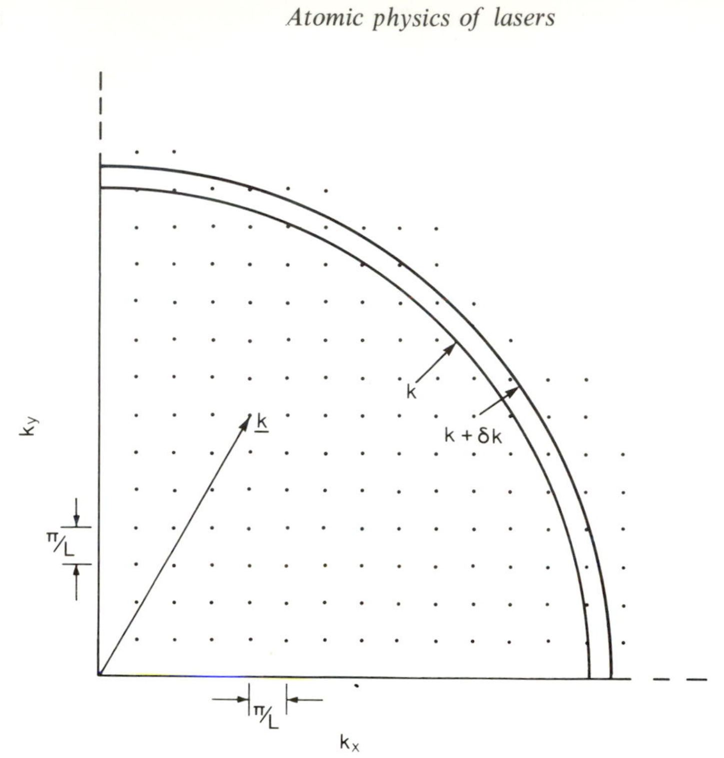

Figure 2.1. Points representing the allowed values of vector $\boldsymbol{k}$, when $k_z = 0$. For $k_z \neq 0$, all the points along the k_x and k_y axis are included except when $k_x = k_y = 0$.

and the number of modes per unit volume of the cavity, $\rho(k)$, in the interval δk, is written

$$\rho(k)\,\delta k = (k^2/\pi^2) \tag{2.4}$$

An alternative expression to (2.4), for the density of modes in frequency space is

$$\rho(v)\,dv = (8\pi v^2/c^3)\,dv \tag{2.5}$$

where we have used the familiar relation $k = 2\pi v/c$.

Having derived the modes of the electromagnetic field we are now in a position to find the energy, if the amplitudes of the modes can be calculated by some other means. The electric fields must satisfy the wave equation

$$\nabla^2 \boldsymbol{E} = \frac{1}{c^2}\frac{\partial^2 \boldsymbol{E}}{\partial t^2} \tag{2.6}$$

where c is the velocity of light. Substituting the values given by (2.1) into the equation we find

$$\frac{\partial^2 \boldsymbol{E}(t)}{\partial t^2} = -4\pi^2\nu^2\boldsymbol{E}(t) \tag{2.7}$$

where the relation, $2\pi\nu = ck$, has been used. The amplitudes are therefore given by the simple expression

$$\boldsymbol{E}(t) = \boldsymbol{E}_0 \exp(-i2\pi\nu t) \tag{2.8}$$

Furthermore the total energy in the cavity box is related to the electric field vector, $\boldsymbol{E}$, and the magnetic field vector, $\boldsymbol{B}$, by the classical relation

$$E_T = \frac{1}{2}\int_V (\varepsilon_0 \boldsymbol{E}^2 + \boldsymbol{B}^2/\mu_0^2)\, dV \tag{2.9}$$

where the integration extends over the volume of the box, and ε_0 and μ_0 are the electric permittivity and magnetic permeability of free space. The two terms in Equation (2.9) have the same magnitude, and so the time average over one cycle of the field for the energy in a mode is

$$\bar{E}_{\nu T} = \frac{1}{2}\int_V \varepsilon_0 |\boldsymbol{E}|^2\, dV \tag{2.10}$$

Accordingly, each mode with an amplitude $|E_0|$ will have an average energy proportional to $|E_0|^2$, and, in classical physics, this amplitude is a continuous variable. In the quantum theory of light, however, the energies in the modes are given by the equation

$$E_\nu = (n+\tfrac{1}{2})h\nu \tag{2.11}$$

where ν is the frequency and n is an integer 0, 1, 2, ... Evidently the field has been quantized by associating each mode with a quantum harmonic oscillator. (Note that Equation (2.7) has the same form as a harmonic oscillator, and this can be quantized in an analogous way to the vibrational states of molecules, discussed in Section 1.14.) Instead of being continuously variable the energy, for any frequency ν, is given in units of $h\nu$, and the quanta are known as photons. The number of photons in any mode is given by the quantum number n.

In thermal equilibrium the atoms (and molecules) of the cavity will be continuously absorbing and emitting photons and the number of photons n, associated with any mode, will be subject to statistical fluctuations. These can be calculated using normal thermodynamic relations, except that any integrals are now replaced by summations since we are now dealing with discrete quantities.

The probability of there being n photons in a particular mode is, according to Maxwell–Boltzmann statistics,

$$P_n = \frac{\exp(-nh\nu/kT)}{\sum_n^{\infty} \exp(-nh\nu/kT)} \tag{2.12}$$

where k is the Boltzmann constant (not to be confused with the modulus of the wavevector $|\boldsymbol{k}|$) and T is the absolute temperature. Here the denominator ensures the normalization condition $\sum_n^{\infty} P_n = 1$. Note that photons, unlike electrons, are bosons and there are no restrictions on the number occupying each mode. Also there are two different modes for each wavevector $\boldsymbol{k}$ (or $\boldsymbol{\nu}$ in frequency space) corresponding to the different directions of $\boldsymbol{E}(t)$ and both will have the same probability distribution. The average energy in the mode is

$$\bar{\varepsilon}_v = \frac{\sum_n^{\infty} nh\nu \exp(-nh\nu/kT)}{\sum \exp(-nh\nu)} = \frac{-d}{d\beta}\left(\log \sum_n^{\infty} \exp(-\beta nh\nu)\right) \tag{2.13}$$

where the abbreviation, $\beta = 1/kT$, has been used. Writing down explicitly the summation on the right hand side of Equation (2.9), the average energy is then

$$\bar{\varepsilon}_v = \frac{-d}{d\beta}\log\left(\frac{1}{1-\exp(-\beta nh\nu)}\right) = \frac{h\nu}{(\exp(h\nu/kT)-1)} \tag{2.14}$$

To calculate the total energy density, the average energy per mode is multiplied by the mode density given by Equation (2.5). Thus, the total energy density in the frequency interval, $d\nu$, is given by the equation

$$W(\nu)\, d\nu = \frac{8\pi\nu^2}{c^3}\frac{h\nu\, d\nu}{(\exp(h\nu/kT)-1)} \tag{2.15}$$

and is known as Planck's radiation law. Further, the outward flow of energy per unit time from any volume within the cavity is given by the surface integral

$$F = \int_s \left(\boldsymbol{E} \times \frac{\boldsymbol{B}}{\mu_0}\right) . ds \tag{2.16}$$

where $\boldsymbol{E} \times \boldsymbol{B}/\mu_0$ is the Poynting vector and S is the surface enclosing the volume, V. This can be related to the total energy, given by (2.9), and it can be shown that, corresponding to Equation (2.15), there is an expression

$$R(\nu)\, d\nu = \frac{W(\nu)c}{4}\, d\nu \tag{2.17}$$

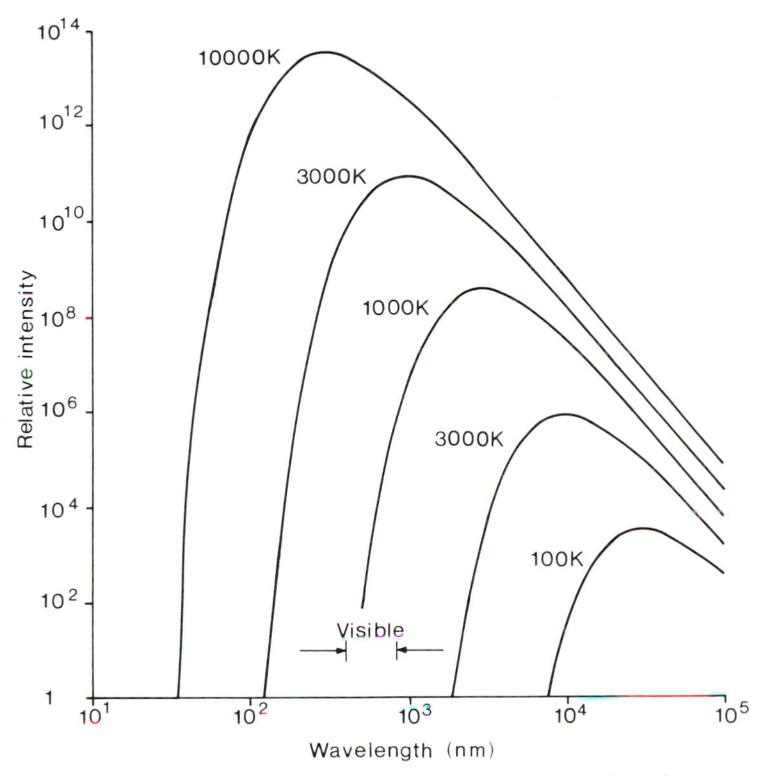

Figure 2.2. Spectral distribution of radiation from a black body at various absolute temperatures. Notice that at long wavelengths the curves approach the Rayleigh-Jeans formula $R_\lambda = 2\pi ckT/\lambda^4$.

for the power emitted by a unit area of a black body between v and $v + dv$. Figure 2.2 shows this for various absolute temperatures.

2.2. Einstein coefficients and stimulated emission

We shall now consider the interaction of radiation with atoms and molecules using a phenomenological theory due to Einstein. Using this theory, Einstein showed that stimulated emission was a necessary consequence of the Planck radiation formula (2.15).

Consider a cavity containing atoms and radiation in equilibrium at a certain temperature. For reasons of simplicity let us assume that the atoms are all of the same type and can exist only in two non-degenerate states 1 and 2, where 1 is the ground state. Radiation from the cavity can interact (neglecting non-resonant processes whose cross-sections are many orders of magnitude smaller), with the atoms if its frequency is in a narrow range centred around ν_{12}, where ν_{12} is given by the familiar equation

$$E_2 - E_1 = h\nu_{12} \tag{2.18}$$

where E_1 and E_2 are the energies of the levels. Three processes are possible:

1. An atom in the excited state 2, can decay by spontaneous emission to the ground state. The probability per unit time of this taking place is denoted by the coefficient A_{21}.

2. An atom in the ground state 1 can absorb radiation from the cavity and be transferred to the excited state. The rate at which this takes place depends not only on the intrinsic nature of the atomic states but also on the radiation density. For a single atom the probability per unit time of a photon being absorbed is $B_{12}\,W(\nu_{12})$. B_{12} is the Einstein coefficient and $W(\nu_{12})$ is the energy density of radiation at frequency ν_{12}. Notice that W is the energy per unit frequency interval, per unit volume.

3. Stimulated emission can take place. A photon of frequency ν_{12} can cause an atom to decay from the excited state 2 to the ground state. The second photon is emitted into the same cavity mode as the original photon 1, and the probability per unit time of this occurring, for a single atom, is $B_{21}W(\nu_{12})$.

If the number of atoms in state 1 is N_1, and the number in 2 is N_2, the rate at which atoms are transferred from state 1 to 2 is

$$\frac{dN_1}{dt} = N_1\,B_{12}\,W(\nu_{12}) \tag{2.19}$$

Similarly, the rate at which atoms are transferred from state 2 to state 1 is

$$\frac{dN_2}{dt} = N_2A_{21} + N_2B_2\,W(\nu_{12}) \tag{2.20}$$

In thermal equilibrium, in an enclosed cavity, these two rates must be equal and we thus obtain the following equation relating the populations of the two states

$$\frac{N_1}{N_2} = \frac{A_{21} + B_{21}\,W(\nu_{12})}{B_{12}\,W(\nu_{12})} \tag{2.21}$$

Also, the thermal equilibrium condition

$$\frac{N_2}{N_1} = \exp((E_1 - E_2)/kT) = \exp(-h\nu_{12}/kT) \tag{2.22}$$

must apply. Equating (2.21) and (2.22) gives an equation for the energy density, in terms of the Einstein coefficients. This is written

$$W(\nu_{12}) = \frac{A_{21}}{B_{12}\exp(h\nu_{12}/kT) - B_{21}} \tag{2.23}$$

and this can be compared with the Planck equation (2.15). Essentially, these are only equivalent if the Einstein coefficients can be related by the equations†

$$A_{21} = \frac{8\pi\nu^3 h}{c^3} B_{21} \tag{2.24}$$

and

$$B_{21} = B_{12}$$

It is also now apparent that the equivalence of Equations (2.23) and (2.15) cannot be established without the existence of stimulated emission. The second Equation (2.24) applies only when the two levels are non-degenerate. If level 1 has a degeneracy g_1, and level 2 has a degeneracy g_2, the equation becomes

$$g_1 B_{12} = g_2 B_{21} \tag{2.25}$$

This can be proved by altering the thermal distribution equation (2.22) to account for the extra energy levels. Thus, the distribution is changed to

$$\frac{g_1 N_2}{g_2 N_1} = \exp(-h\nu_{12}/kT) \tag{2.26}$$

A further point concerns this thermal distribution condition. It can be seen that $g_1 N_2$ (or N_2 for Equation (2.22)) cannot be larger than $g_2 N_1$ (or N_1 for Equation (2.22)), no matter how high the temperature. We shall show in the next chapter that a population inversion, that is when the left hand side of Equation (2.26) is greater than unity, is a necessary condition for the production of laser light. This can be achieved, for some levels, when external influences such as an electrical discharge are present, but it is worth pointing out that the seemingly inviolate nature of Equation (2.26) was one of the stumbling blocks in the history of the laser.

2.3. *Properties of the Einstein coefficients*

It is worthwhile considering the properties of the Einstein coefficients and how far they can be applied in other situations, such as the interaction of a

† Note that the coefficients B_{21} and B_{12} will differ by 2π if $\rho(\omega)$ is used instead of $\rho(\nu)$. In the following sections the angular frequency is used exclusively, and Equation (2.24) is $A_{21} = (\omega^3\hbar/\pi^2c^3)B^{\omega}_{21}$.

parallel light beam with an ensemble of atoms (or molecules). Clearly, these considerations are most important to the theory of lasers.

Firstly, the B coefficients are averaged over all directions of the light beam and atomic orientations. (The atomic orientation is defined relative to its total angular momentum, J.) The detailed microscopic theory, which is given in the next sections, shows that the interaction depends on the orientation of the atoms relative to the light beam. However, in all the lasers considered here, the atoms (or molecules) are randomly orientated (e.g. in a gas laser), and averaging over all the orientations of the atom is equivalent to averaging over both the incident light direction and the orientations of the atom. Of course, if the atoms were orientated in one direction, for example in a solid, the coefficients could not be used directly.

Secondly, the B coefficients refer to averages over a continuous spectrum of frequencies. In the case of a parallel light beam with a very narrow frequency range we need, for example, to consider the effects of Doppler shifts. Some of the atoms may be moving in a direction, and with a velocity, such that their interaction with the light beam is much less probable, and for this case the coefficients cannot be used by themselves. In fact the interaction depends on the homogeneous and inhomogeneous broadening of the transition, and these effects are considered in detail at the end of this chapter. In Chapter 3 we show how the B coefficients can still be retained as long as they are duly scaled to allow for these broadening effects.

Thirdly, the directional properties of the radiation need to be considered. For a parallel incident beam the spontaneously emitted radiation can be in many directions. Photons are effectively scattered out of the incident beam by this process. For randomly polarized incident light the angular distribution of the radiation is isotropic, that is, there is no preferred direction. In contrast to this, the stimulated emission is in the same direction as the incident radiation. Thus the extra photon goes into the same cavity mode as the original photon. Note that Equation (2.23) does not show this directly since it refers to only the magnitude of the wavevector. This equation would still hold if equal numbers of photons were scattered out of a particular mode (of a given frequency), as were scattered into it. Stimulated emission leads to a general increase in the intensity of the beam whilst spontaneous emission reduces it. Also the phase of the incident beam is preserved and this gives laser beams an exceptionally high degree of spatial and temporal coherence. In Section 2.10 the quantum electrodynamic equations which relate the directions of the radiation are given.

The directional properties of stimulated emission can be understood on a classical basis by considering the radiation from oscillating dipoles. Consider a plane wave propagating along the z axis with its electric field linearly polarized in the x direction, and a free dipole (undamped) at the origin of the co-ordinates, which can oscillate in the x direction, at the frequency of the incident wave. A simple calculation shows that, if the phase angle between the dipole oscillations and the wave is $\pi/2$, then energy is extracted from the oscillator by the field, which corresponds to stimulated emission, whilst, for a phase angle of $-\pi/2$, the field

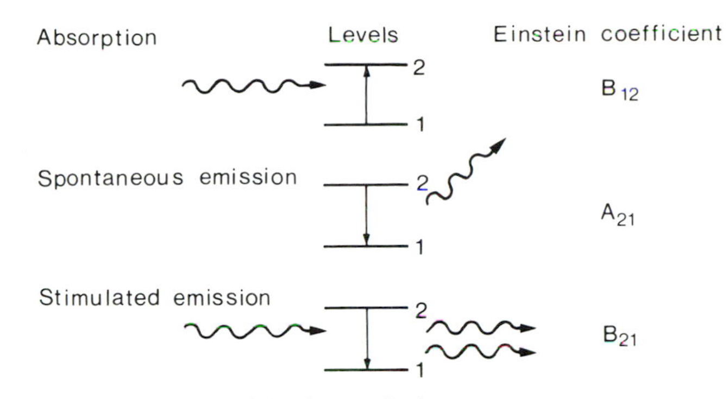

Figure 2.3. Schematic diagram of the three radiative processes.

does work on the dipole, corresponding to absorption. The total energy flow in the radiation field at any point can be calculated by first adding vectorially the field ($\boldsymbol{E}$ and $\boldsymbol{B}$) from the dipole to that in the plane wave and then calculating the Poynting vector. Because of the interference between the scattered dipole wave and the plane wave there is only a net increase (or decrease for absorption) in energy flow along the z axis at points where $\theta < (\lambda/z)^{1/2}$, θ being the angle between the z axis and the position vector $\boldsymbol{r}$. Away from the axis the interference terms give rise to very rapid oscillations with respect to θ, which average out over any small area of detection to give the same intensity as in the incident beam. Now a more realistic picture of stimulated emission (in a laser) would involve a sheet of dipoles in the xy plane all oscillating with a common phase. It is fairly obvious that for an infinite sheet the dipole emission must be only in the z direction. (This is just Huygens' principle in optics and, indeed, we can expect that for a finite sheet of radial dimension, a, then the beam will be contained in a half angle $\theta \simeq \lambda/a$). A careful calculation of the phases shows that for an infinite sheet of dipoles the contribution from all the dipoles is shifted by $\pi/2$ from that of a single dipole so that the stimulated emission has exactly the same phase and optical dependence as the incident wave. For spontaneous emission the phases of the dipoles will be randomly distributed so radiation will not be confined to just the z axis.

Figure 2.3 illustrates the three processes for an incident parallel light beam.

2.4. *Time-dependent wave equation*

In Chapter 1, the stationary states of atoms and molecules were calculated using the time-independent Schrödinger equation. For calculations requiring the rate at which (microscopic) processes taken place we need to consider the time-dependent equation. The transition rates, which up to now have been described by the A and B coefficients, can then be enunciated in terms of the microscopic properties of the atoms.

Consider, as previously, an atom absorbing light from a parallel beam so that it undergoes transitions from its ground state to an excited state. The wave equation describing the interaction is

$$H\psi = i\hbar\frac{\partial\psi}{\partial t} \tag{2.27}$$

where H is the Hamiltonian operator, and is composed of two terms, the original interaction within the atom and the extra interaction from the incident light beam. H can thus be written

$$H_0 + \Delta H \tag{2.28}$$

where H_0 is the Hamiltonian of the atom by itself. Further, the atomic wave function can be written as a sum of two terms, ϕ_1 and ϕ_2 representing the stationary states of the ground and excited atomic levels respectively. These wave functions have the familiar time dependence of stationary states and are written

$$\phi_1 = \phi_{10}(\boldsymbol{r})\exp(-iE_1t/\hbar) \tag{2.29}$$

and

$$\phi_2 = \phi_{20}(\boldsymbol{r})\exp(-iE_2t/\hbar)$$

where E_1 and E_2 are the energies of the atomic states. The total wave function is

$$\psi = C_1(t)\phi_1 + C_2(t)\phi_2 \tag{2.30}$$

where $C_1(t)$ and $C_2(t)$ represent the amplitudes of the two components. Evidently, $|C_1(t)|^2$ represents the probability of finding the atom in state 1, and $|C_2(t)|^2$ of finding the atom in state 2. The number of atoms in state 1 is therefore $N_0|C_1(t)|^2$ and in state 2, $N_0|C_2(t)|^2$, where N_0 is the original number of atoms. Conservation requires that

$$|C_1(t)|^2 + |C_2(t)|^2 = 1 \tag{2.31}$$

which is equivalent to demanding that the integral $\int \psi^*\psi\, dv$, over all space, is unity. The rate at which atoms are transferred from level 1 to level 2 can be found by substituting the wave function (2.30) into the time-dependent wave equation (2.27). Thus

$$(H_0 + \Delta H)(C_1\phi_1 + C_2\phi_2) = i\hbar\frac{\partial}{\partial t}(C_1(t)\phi_1 + C_2(t)\phi_2) \tag{2.32}$$

and this reduces to the equation

$$\Delta H(C_1\phi_1 + C_2\phi_2) = i\hbar\left(\phi_1\frac{\partial C_1(t)}{\partial t} + \phi_2\frac{\partial C_2(t)}{\partial t}\right) \tag{2.33}$$

where the following relations have been used

$$H_0\phi_1 = i\hbar\frac{\partial\phi_1}{\partial t} \qquad \text{and} \qquad H_0\phi_2 = i\hbar\frac{\partial\phi_2}{\partial t} \tag{2.34}$$

Two equations relating the time dependence of the coefficients $C_1(t)$ and $C_2(t)$ can be obtained from Equation (2.33). Firstly, operating on both sides of the equation with ϕ_2^* (complex conjugate of ϕ_2) and integrating over all space, gives

$$C_1(t)\int\phi_2^*\Delta H\phi_1\,dv + C_2(t)\int\phi_2^*\Delta H\phi_2\,dv = i\hbar\frac{\partial C_2(t)}{\partial t} \tag{2.35}$$

and similarly, operating with ϕ_1^* gives

$$C_1(t)\int\phi_1^*\Delta H\phi_1\,dv + C_2(t)\int\phi_1^*\Delta H\phi_2\,dv = i\hbar\frac{\partial C_1(t)}{\partial t} \tag{2.36}$$

Substituting the explicit wave functions (2.29) for ϕ_1 and ϕ_2 gives the two equations

$$C_1(t)\exp(i\omega_{12}t)H'_{21} + C_2(t)H'_{22} = i\hbar\frac{\partial C_2(t)}{\partial t} \tag{2.37}$$

and

$$C_1(t)H'_{11} + C_2(t)\exp(-i\omega_{12}t)H'_{12} = i\hbar\frac{\partial C_1(t)}{\partial t} \tag{2.38}$$

where the matrix elements, H'_{ij}, are

$$H'_{jk} = \int\phi_{j0}^*(\boldsymbol{r})\Delta H\phi_{k0}(\boldsymbol{r})\,dv \tag{2.39}$$

and

$$\hbar\omega_{12} = E_2 - E_1$$

In as much as Equations (2.37) and (2.38) involve only two levels they are exact. Despite their apparent simplicity, exact solutions are only possible for certain restricted cases of the operator ΔH. Before we go on to consider the form of ΔH, in the next section, a more general equation which applies to systems of many levels, can be written

$$\dot{C}_s(t) = \frac{1}{i\hbar} \sum C_n(t) H'_{sn} \exp(i\omega_{sn} t) \tag{2.40}$$

where $\dot{C}_s(t) \equiv \partial/\partial t(C_s(t))$ and H'_{sn} is defined by Equation (2.39). Here the angular frequency ω_{sn} is defined by the equation

$$\hbar\omega_{sn} = E_s - E_n \tag{2.41}$$

where E_s and E_n are the energies of the levels s and n, for the original Hamiltonian H_0. Evidently, Equation (2.40) reduces to the two coupled Equations (2.37) and (2.38) when only two levels are involved.

Approximate solutions to Equation (2.40) can be obtained, when ΔH is much smaller than H_0, using time-dependent perturbation theory.† The first approximation to the solution is obtained by substituting $C_n(0)$ for $C_n(t)$ on the right hand side of Equation (2.40) and integrating from 0 to t. Thus

$$C_s(t) = \frac{1}{i\hbar} C_s(0) \int_0^t H'_{ss}\, dt + \frac{1}{i\hbar} \sum_n C_n(0) \int_0^t H'_{sn} \exp(i\omega_{sn} t)\, dt \tag{2.42}$$

If the system is in a definite state n, at time zero, when the perturbation is 'switched on', then $C_n(0) = 1$, and all the other coefficients on the right hand side of (2.41) are zero. The first-order time dependence of the amplitude, C_s, is therefore given by the equation

$$C(t) = \frac{1}{i\hbar} \int_0^t H'_{sn} \exp(i\omega_{sn} t)\, dt \tag{2.43}$$

which applies as long as $C_s(t) \ll 1$. Further, if the perturbing Hamiltonian ΔH is time independent, then Equation (2.43) can be integrated directly to give

$$C_s(t) = -\frac{H'_{sn}}{\hbar\omega_{sn}} (\exp(i\omega_{sn} t) - 1) \tag{2.44}$$

The probability that an atom is found in a state s, after a certain time t, is therefore given by the equation

$$P_{sn}(t) = |C_s(t)|^2 \tag{2.45}$$

---

† See for example, Schiff, L.G., 1948, *Quantum Mechanics* (McGraw-Hill, New York).

and, substituting Equation (2.44) into this gives

$$P_{sn}(t) = 4|H_{sn}|^2 \frac{\sin^2(\omega_{sn}t/2)}{\hbar^2\omega_{sn}^2} \tag{2.46}$$

This function is plotted in Figure 2.4. When ω_{sn} tends to zero this reduces to

$$\underset{\omega_{sn}\to 0}{P_{sn}(t)} = \frac{|H_{sn}|^2 t^2}{\hbar^2} \tag{2.47}$$

so that the transition probability increases proportional to t^2. When ω_{sn} is not zero the function oscillates between zero and $(2/\omega_{sn}^2)$ with a frequency $\omega_{sn}/2\pi$, so that the average probability is

$$\overline{P_{sn}(t)} = \frac{2|H'_{sn}|^2}{|E_s - E_n|^2} \tag{2.48}$$

An important case is one where the level s forms a continuum of states such as is encountered in the emission of light. Although there are only two atomic states, the photon can be in a large number of modes whose density is given by Equation (2.5). The total probability of a transition to the final states is then given by the sum

$$P_{sn}(t) = \sum_s |C_s(t)|^2 = 4|\bar{H}_{sn}|^2 \sum_s \frac{\sin^2(\omega_{sn}t/2)}{\hbar^2\omega_{sn}^2} \tag{2.49}$$

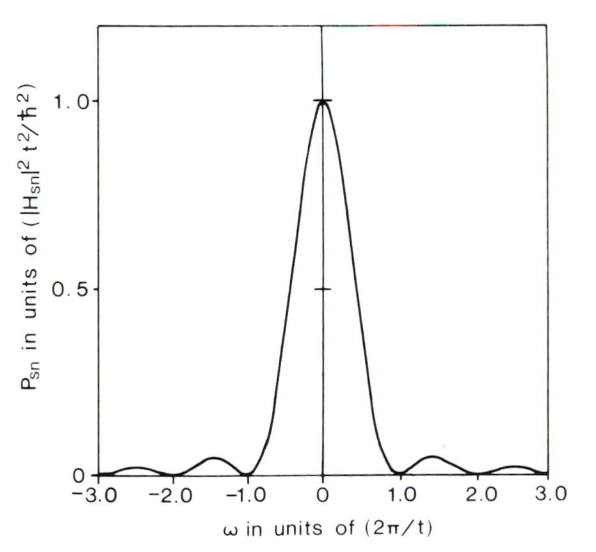

Figure 2.4. The probability function according to Equation (2.46), plotted as a function of ω_{sn} for a fixed t.

where $|\bar{H}_{sn}|^2$ now represents a suitable average value. Assuming that the states form a continuum in a narrow interval about the level s with an energy density $\rho(E_s)$ such that $\rho(E_s)\delta E_s$ is the number of states in the interval from E to $E_s + \delta E_s$, the summation in Equation (2.49) is replaced by an integral, and we obtain

$$P_{sn}(t) = 4|\bar{H}_{sn}|^2\hbar \int_{\hbar(\omega_{sn} - \delta\omega_{sn})}^{\hbar(\omega_{sn} + \delta\omega_{sn})} \rho(E_s) \frac{\sin^2(\omega_{sn}t/2)}{\hbar^2\omega_{sn}^2} \, d\omega_{sn} \tag{2.50}$$

Since $\rho(E_s)$ is constant over this interval it can be removed from the integral. For t larger than $2\pi/\delta\omega_{sn}$ the major contribution to the integral is when ω_{sn} is close to zero, and the range of the integral can be extended from $-\infty$ to $+\infty$. Evaluating this integral gives

$$P_{sn}(t) = \frac{2\pi}{\hbar}|\bar{H}_{sn}|^2\rho(E_s)t \tag{2.51}$$

for the probability function, and the rate of the transition (the probability per unit time of a transition taking place) which is defined as $(\partial P_{sn}/\partial t)$, is

$$\frac{\partial P_{sn}}{\partial t} = \frac{2\pi}{\hbar}|\bar{H}_{sn}|^2\rho(E_s) \tag{2.52}$$

This formula is often called Fermi's 'golden rule no. 2' and was first derived by Dirac! Since this formula applies only for the case of a time-independent perturbation†, it cannot strictly be applied to the absorption and emission of radiation. However, the connection between emission and absorption of radiation as given by the Einstein Equations (2.24) is already evident in Equation (2.52). Here the density of states for absorption is unity (since the incident photon is in a single mode), whereas in spontaneous emission $\rho(E_s)$ is given by the standard Equation (2.5), which now refers to the different modes which the emitted photon can occupy.

In a rigorous quantum electrodynamic (QED) treatment of the absorption emission of radiation the relevant matrix elements are calculated using eigenstates made up from a product of the atomic eigenstates and those of the quantized radiation field. In this way an equation which looks similar to (2.52) can be derived, but which includes both spontaneous and stimulated emission (see Equation (2.193)). However, it is possible to use a semi-classical approach for calculating the absorption. For absorption involving two atomic states we can simply use an equation like (2.52) with the interaction ΔH taken to be the classical energy of an atom in an electromagnetic field. Notice here that density of states factor will be unity. From the absorption rates it is then possible to calculate the

† It can be made quite general for oscillating perturbations as the next section shows.

emission using the Einstein equation (2.24). There are serious drawbacks in this procedure, not the least of which is its inability to give a satisfactory explanation for spontaneous emission. In fact, spontaneous emission arises because of the interaction of the atom with the zero-point oscillations of the electromagnetic field; a concept which does not arise until the field is quantized. This idea was mentioned previously in Section 2.1 and Equation (2.7). For molecular vibrations, the zero-point energy can be verified by direct measurements of the energy levels in different molecules. In the case of the electromagnetic field the most obvious manifestation of the zero-point oscillations is spontaneous emission but they also have a small effect on the stationary atomic energy levels. Thus, the Lamb shift (see Section 1.4) is caused by the different strengths of interaction between the zero-point field and the atomic states. A second drawback concerns stimulated emission. Here the photons are emitted into the same cavity mode as the incident photons, a fact which the black body analysis does not require but one which is adequately taken care of by quantum electrodynamics. However, this will be explained without QED in Section 2.8.

In the next sections we consider the semi-classical theory of the interaction between electromagnetic radiation and atoms. Within this framework, a few of the most important analogous equations of QED are included. This should give the reader a small insight into the basic ideas associated with quantization of the radiation field.

2.5. *The Hamiltonian for an atom in an electromagnetic field*

The Hamiltonian used in Equation (2.28) will now be derived for the case of an atom in an electromagnetic field. In the semi-classical approximation it is assumed that the electric and magnetic fields are continuous variables which are solutions of the Maxwell equations

$$\varepsilon_0 \nabla . \boldsymbol{E} = \rho \tag{2.53}$$

$$\nabla . \boldsymbol{B} = 0 \tag{2.54}$$

$$\nabla \times \boldsymbol{E} = -\frac{\partial \boldsymbol{B}}{\partial t} \tag{2.55}$$

and

$$(1/\mu_0)\nabla \times \boldsymbol{B} = \boldsymbol{J} + \varepsilon_0 \frac{\partial \boldsymbol{E}}{\partial t} \tag{2.56}$$

where $\boldsymbol{J}$ and ρ are the current and charge densities. Here it is assumed that $\varepsilon = \mu = 1$. From these equations it is easy to derive the wave equations for $\boldsymbol{E}$ and

$\boldsymbol{B}$. These are given by the expressions

$$\nabla^2 \boldsymbol{E} = -\mu_0 \varepsilon_0 \frac{\partial^2 \boldsymbol{E}}{\partial t^2} \tag{2.57}$$

and

$$\nabla^2 \boldsymbol{B} = -\mu_0 \varepsilon_0 \frac{\partial^2 \boldsymbol{B}}{\partial t^2} \tag{2.58}$$

In Section 2.1 a solution of (2.57) was given for the standing wave radiation in a closed box.

A more convenient expression for the Hamiltonian can be obtained using the vector and scalar potentials, $\boldsymbol{A}$ and ϕ, which are defined by the equations

$$\boldsymbol{B} = \nabla \times \boldsymbol{A} \tag{2.59}$$

and

$$\boldsymbol{E} = -\nabla \phi - \frac{\partial \boldsymbol{A}}{\partial t} \tag{2.60}$$

Evidently, with these definitions, $\boldsymbol{B}$ and $\boldsymbol{E}$ satisfy Equations (2.54) and (2.55) respectively. An equation relating $\boldsymbol{A}$ and ϕ to the current and charge densities can be obtained by substituting (2.59) and (2.60) into Equations (2.56) and (2.55). The vector and scalar potentials which satisfy Maxwell's equations are therefore given by the equations

$$\nabla \times \nabla \times \boldsymbol{A} + \frac{1}{c^2}\left(\frac{\partial^2 \boldsymbol{A}}{\partial t^2}\right) + \frac{1}{c^2}\frac{\partial}{\partial t}(\nabla \phi) = \mu_0 \boldsymbol{J} \tag{2.61}$$

and

$$-\nabla^2 \phi - \nabla . \frac{\partial \boldsymbol{A}}{\partial t} = \frac{\rho}{\varepsilon_0} \tag{2.62}$$

where the relation $c^2 = 1/\varepsilon_0 \mu_0$, has been used. Thus, in principle, $\boldsymbol{A}$ and ϕ can be obtained if the current and charge densities are known, and from these the electric and magnetic fields can be found from Equations (2.59) and (2.60). However, the Equations (2.61) and (2.62) can be considerably simplified by means of gauge transformation. Consider the transformation

$$\boldsymbol{A}' = \boldsymbol{A} + \nabla \chi \tag{2.63}$$

and

$$\phi' = \phi - \frac{\partial \chi}{\partial t} \tag{2.64}$$

where χ is an arbitrary function of $\boldsymbol{r}$ and t. Since any scalar function satisfies the relation, $\nabla \times \nabla\chi = 0$, both $\boldsymbol{E}$ and $\boldsymbol{B}$ are unchanged by this transformation. We are now free to choose χ in a manner which most simplifies Equations (2.61) and (2.62). A convenient transformation is the Coulomb gauge, in which the vector potential $\boldsymbol{A}$ satisfies the condition $\nabla.\boldsymbol{A} = 0$. Substituting this condition into (2.63) gives the equation

$$\nabla^2\chi = \nabla.\boldsymbol{A} \tag{2.64a}$$

for the necessary restriction on χ imposed by this transformation. With this choice of gauge, Equations (2.61) and (2.62) are simplified to

$$-\nabla^2\boldsymbol{A} + \frac{1}{c^2}\frac{\partial^2\boldsymbol{A}}{\partial t^2} + \frac{1}{c^2}\frac{\partial}{\partial t}(\nabla\phi) = \mu_0\boldsymbol{J} \tag{2.65}$$

and

$$-\nabla^2\phi = \frac{\rho}{\varepsilon_0} \tag{2.66}$$

The term Coulomb gauge is now obvious, since the scalar potential ϕ is now a solution of Poisson's equation. ϕ, for a given time, is therefore the same as the potential from a system of static charges, and can be written

$$\phi(r) = \frac{1}{4\pi\varepsilon_0}\int\frac{\rho(r')}{(r-r')}\,dv' \tag{2.67}$$

(An alternative transformation, which gives symmetrical equations for $\boldsymbol{A}$ and ϕ, can be obtained using the Lorentz condition $\nabla.\boldsymbol{A} + (1/c)(\partial\phi/\partial t) = 0$. This can be immediately substituted into Equations (2.61) and (2.62) to obtain two uncoupled equations for $\boldsymbol{A}$ and ϕ. The equation for the scalar function is found by inserting Equations (2.63) and (2.64) into the Lorentz equations.)

A further reduction of Equations (2.65) is possible if the current density, $\boldsymbol{J}$, is separated into two components, one with zero divergence and one with zero curl. Thus

$$\boldsymbol{J} = \boldsymbol{J}_T + \boldsymbol{J}_L \tag{2.68}$$

where

$$\nabla.\boldsymbol{J}_T = 0 \tag{2.69}$$

$$\nabla \times \boldsymbol{J}_L = 0 \tag{2.70}$$

and the continuity equation relating $\boldsymbol{J}$ and ρ, is therefore

$$\nabla . \boldsymbol{J}_L = -\frac{d\rho}{dt} \tag{2.71}$$

Any vector which has zero curl can be expressed in terms of a scalar field ψ, by the equation

$$\boldsymbol{J}_L = \nabla\psi \tag{2.72}$$

and hence, using (2.70), we obtain

$$\nabla^2\psi = -\frac{d\rho}{dt} \tag{2.73}$$

From (2.6) the scalar field, ψ, is therefore

$$\psi = \varepsilon_0 \frac{\partial\phi}{\partial t} \tag{2.74}$$

and substituting this into Equation (2.72) gives

$$\boldsymbol{J}_L = \varepsilon_0 \nabla \frac{\partial\phi}{\partial t} \tag{2.75}$$

Equation (2.65) is therefore simplified to one containing only the vector potential and this is written

$$-\nabla^2 \boldsymbol{A} + \frac{1}{c^2}\frac{\partial^2 \boldsymbol{A}}{\partial t^2} = \mu_0 \boldsymbol{J}_T \tag{2.76}$$

The general solution to this is

$$A(r, t) = \frac{\mu_0}{4\pi}\int \frac{\boldsymbol{J}_T(\boldsymbol{r}', t')\, dv'}{|\boldsymbol{r}-\boldsymbol{r}'|} \tag{2.77}$$

where

$$t' = t - \frac{|\boldsymbol{r}-\boldsymbol{r}'|}{c} \tag{2.78}$$

Note that the calculation of $\boldsymbol{A}$ at any time t requires a knowledge of the current distribution at the retarded time t', so the equation allows for the propagation of the vector potential, $\boldsymbol{A}$.

The electric and magnetic fields can be calculated from Equation (2.59) and (2.60) if the vector and scalar potentials are known. However, the scalar potential as determined from (2.66) cannot be related to the fields in an electromagnetic wave. (The left-hand side of Equation (2.66) would require terms like $\partial^2\phi/\partial t^2$ for this to be untrue. For the Lorentz gauge, however, the field in the oscillating wave is related to ϕ.) Formally, we can separate the electric field into two components which are related by the following equations

$$\boldsymbol{E} = \boldsymbol{E}_T + \boldsymbol{E}_L \tag{2.79}$$

and

$$\boldsymbol{E}_L = -\nabla\phi \tag{2.80}$$

and

$$\boldsymbol{E}_T = -\frac{\partial \boldsymbol{A}}{\partial t} \tag{2.81}$$

The longitudinal component, E_L, is just related to the charge distribution, ρ, by Poisson's equation, whereas the transverse component is associated with the field in the electromagnetic wave. Thus, $\boldsymbol{E}_T$ is the electric field which appears in solutions of Equation (2.57). $\boldsymbol{B}$, of course, is always transverse according to the Maxwell equation (2.54).

The Hamiltonian can now be calculated using the vector and scalar potentials, $\boldsymbol{A}$ and ϕ. Firstly, consider an electron moving in a combined magnetic and electric field. The (Lorentz) force acting on this electron is

$$\boldsymbol{F} = q(\boldsymbol{E}+\boldsymbol{u}\times\boldsymbol{B}) \tag{2.82}$$

where q is the charge ($=-e$ for the electron) and $\boldsymbol{u}$ is the velocity. Using Equations (2.59) and (2.60) this can be written

$$\boldsymbol{F} = q\left(-\nabla\phi-\frac{\partial \boldsymbol{A}}{\partial t}+\boldsymbol{V}\times\nabla\times\boldsymbol{A}\right) \tag{2.83}$$

which upon expanding the last term and noting that

$$\frac{d\boldsymbol{A}}{dt} = \frac{\partial \boldsymbol{A}}{\partial t}+(\boldsymbol{u}\,.\,\nabla)\boldsymbol{A} \tag{2.84}$$

gives

$$\boldsymbol{F} = q\left(-\nabla(\phi-\boldsymbol{u}\,.\,\boldsymbol{A})-\frac{\mathrm{d}\boldsymbol{A}}{\mathrm{d}t}\right) \tag{2.85}$$

From this it can be shown† that the proper Lagrangian is

$$L = T - q\phi + q\boldsymbol{u}\,.\,\boldsymbol{A} \tag{2.86}$$

where T is the kinetic energy of the electron. The Hamiltonian is therefore

$$H_e = \frac{(\boldsymbol{p} - q\boldsymbol{A})^2}{2m} + q\phi \tag{2.87}$$

where $\boldsymbol{p}$ is the electron momentum. Equation (2.87) can be expanded to give

$$H_e = \frac{\boldsymbol{p}^2}{2m} + q\phi - \frac{q}{2m}(\boldsymbol{p}\,.\,\boldsymbol{A} + \boldsymbol{A}\,.\,\boldsymbol{p}) + \frac{q^2}{2m}|\boldsymbol{A}|^2 \tag{2.88}$$

Now, in the Coulomb gauge, the first part of this equation is the Hamiltonian of an electron in the static field of the nucleus. The extra term for Equation (2.28) is therefore

$$\Delta H = -\frac{q}{2m}(\boldsymbol{p}\,.\,\boldsymbol{A} + \boldsymbol{A}\,.\,\boldsymbol{p}) + \frac{q^2}{2m}|\boldsymbol{A}|^2 \tag{2.89}$$

and because $\boldsymbol{\nabla}\,.\,\boldsymbol{A} = 0$, this can be simplified to

$$\Delta H = -\frac{q}{m}\boldsymbol{A}\,.\,\boldsymbol{p} + \frac{q^2}{2m}|\boldsymbol{A}|^2 \tag{2.90}$$

The second term is much smaller than the first term and can often be neglected. In QED, this is associated with two-photon processes which are generally much less probable unless the allowed transitions are forbidden by selection rules. Equation (2.88) can be generalized to an atom by summing over all the electrons. For wavelengths much larger than atomic dimensions, this is achieved by replacing the momentum $\boldsymbol{p}$, by the total momentum $\boldsymbol{P}$. Thus, Equation (2.88) is finally reduced to

$$\Delta H = -\frac{q}{m}\boldsymbol{A}\,.\,\boldsymbol{P} \tag{2.91}$$

where

$$\boldsymbol{P} = \sum_j \boldsymbol{p}_j \tag{2.92}$$

and the summation extends over all the electrons. The vector potential $\boldsymbol{A}$ is a continuous variable and it cannot therefore be used to calculate any atomic

† See for example, Schiff, L.G., 1948, *Quantum Mechanics* (McGraw-Hill, New York), p. 133.

emission processes. Here the concern is with small numbers of photons; for spontaneous emission only one photon is involved (neglecting the $|\boldsymbol{A}|^2$ term), whilst in stimulated emission two photons take part. $\boldsymbol{A}$, as derived from the classical field, can only be representative for a large number of photons.

2.6. *Properties of an electromagnetic plane wave*

Before calculating the matrix elements for absorption it is worthwhile reviewing the properties of an electromagnetic plane wave. The vector potential of a travelling wave is given by the equation

$$\boldsymbol{A} = \frac{A_0}{2}\hat{\boldsymbol{\varepsilon}}(\exp(i\boldsymbol{k}\,.\,\boldsymbol{r} - i\omega t + \delta) + \exp(-i\boldsymbol{k}\,.\,\boldsymbol{r} + i\omega t + \delta)) \tag{2.93}$$

where $k = \omega/c$, and $\hat{\boldsymbol{\varepsilon}}$ is the unit vector representing the polarization. (This is a linearly polarized wave, the extension to any state of polarization can be achieved by summing any two independent linearly polarized waves, having in general different phases, $\delta(\omega)$.) For the Coulomb gauge ($\boldsymbol{\nabla}\,.\,\boldsymbol{A} = 0$), $\boldsymbol{A}$ is orthogonal to the direction of propagation, that is

$$\boldsymbol{k}\,.\,\boldsymbol{A} = 0 \tag{2.94}$$

and the wave is transverse. This was discussed in the previous section. From Equations (2.59) and (2.55) the electric and magnetic fields, corresponding to the vector potential (2.93), are

$$\boldsymbol{E} = -\omega A_0 \hat{\boldsymbol{\varepsilon}} \sin(\boldsymbol{k}\,.\,\boldsymbol{r} - \omega t + \delta) \tag{2.95}$$

and

$$\boldsymbol{B} = -A_0 \boldsymbol{k} \times \hat{\boldsymbol{\varepsilon}} \sin(\boldsymbol{k}\,.\,\boldsymbol{r} - \omega t + \delta) \tag{2.96}$$

$\boldsymbol{E}$ is in the same direction as $\boldsymbol{A}$, but $\boldsymbol{B}$ is perpendicular to $\boldsymbol{A}$. All three vectors $\boldsymbol{A}$, $\boldsymbol{E}$ and $\boldsymbol{B}$ are transverse. These are shown in Figure 2.5. A formula for the total energy† density can be obtained using Equation (2.9) and taking the mean value over one cycle of the field. Thus

$$W_\omega = \tfrac{1}{2}\varepsilon_0 \omega^2 A_0^2 \tag{2.97}$$

and this must be equal to $N\hbar\omega/V$, where N is the number of photons with

† This is not the energy per unit frequency interval, since we are dealing with a monochromatic wave.

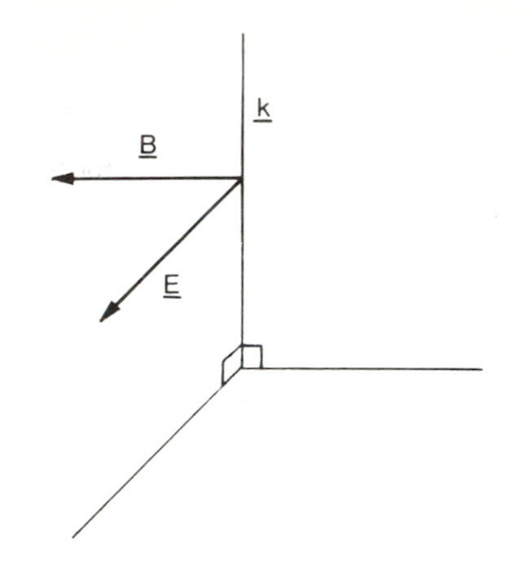

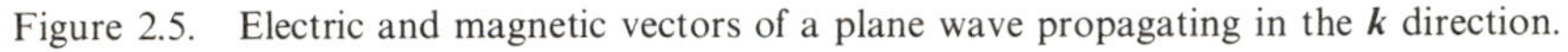
Figure 2.5. Electric and magnetic vectors of a plane wave propagating in the $\boldsymbol{k}$ direction.

frequency ω. (Note that N must be large.) The amplitude can therefore be related to the number of photons by the expression

$$A_0^2 = \frac{2N\hbar\omega}{\varepsilon_0\omega^2 V} \tag{2.98}$$

where V is the volume in which the photons are counted. The flux density or intensity of the wave, which is just the Poynting vector ($\boldsymbol{E}\times\boldsymbol{B}/\mu_0$) averaged over one field cycle, is

$$I_\omega = \tfrac{1}{2}\varepsilon_0\omega^2 A_0^2 c = \tfrac{1}{2}\varepsilon_0 E_0^2 c$$

and, using (2.98), this gives

$$I_\omega = \frac{N\hbar\omega c}{V} \tag{2.100}$$

Note, also, that the continuity equation, $I = Wc$, is obeyed.

2.7. *Transition probability for absorption*

The transition rates for absorption can now be calculated using the Hamiltonian (2.91) and the coupled Equations (2.37) and (2.38). For the plane wave, we incorporate the phases into the amplitudes, and write

$$\boldsymbol{A} = \frac{A_0}{2}\hat{\boldsymbol{\varepsilon}}(\exp(i\boldsymbol{k}.\boldsymbol{r}-\omega t)+\exp(-i\boldsymbol{k}.\boldsymbol{r}+\omega t)) \tag{2.101}$$

where now A_0 can have real and complex components. Initially, we consider $\hat{\boldsymbol{\varepsilon}}$ is fixed in space and so represents a plane wave.

An analogous expression to (2.88) in QED is

$$\Delta H = \frac{-q}{m}\sum_i \hat{\boldsymbol{A}}(\boldsymbol{r}_i).\boldsymbol{p}_i + \frac{q^2}{2m}\sum_i |\hat{\boldsymbol{A}}(\boldsymbol{r}_i)|^2 \tag{2.102}$$

where, as before, $\hat{\boldsymbol{A}}(\boldsymbol{r}_i)$ can be reduced to $\hat{\boldsymbol{A}}(\boldsymbol{r})$ in the long wavelength limit. $\hat{\boldsymbol{A}}(\boldsymbol{r})$ is the quantized vector potential and is given by the expression

$$\hat{\boldsymbol{A}}(\boldsymbol{r}) = \frac{1}{2}\sum_k (\hbar/2\varepsilon_0 V\omega_k)^{1/2}\boldsymbol{\varepsilon}_k(\hat{a}_k \exp(i\boldsymbol{k}.\boldsymbol{r} - i\omega_k t) + \hat{a}_k^+ \exp(-i\boldsymbol{k}.\boldsymbol{r} + i\omega_k t)) \tag{2.103}$$

where $\boldsymbol{\varepsilon}_k$ is the unit vector representing the polarization in the mode $\boldsymbol{k}$. This is a formal representation of the mode structure discussed in Section 2.1 *except* that now the vector potential represents travelling rather than standing waves. Quantization is achieved by using the creation and annihilation operators $\hat{a}_k$ and $\hat{a}_k^+$. $\hat{a}_k^+$ creates a photon of energy $h\omega_k$ in the mode $\boldsymbol{k}$, and $\hat{a}_k$ destroys it. Note there are two modes for each $\boldsymbol{k}$ with different polarizations. These are not distinguished here. The total energy in the field is

$$E_T = \sum_k (n_k + \tfrac{1}{2})h\omega_k \tag{2.104}$$

where n_k is given by the equation

$$\hat{a}_k^+ a_k|n_k\rangle = n_k|n_k\rangle \tag{2.105}$$

Evidently, n_k is the total number of photons in the mode whose wavevector is $\boldsymbol{k}$. Similar expressions to (2.103) can be written for the electric and magnetic fields.

A further simplification of the matrix elements (2.39) is possible. The spatial part of these matrix elements will be proportional to the elements A_{jk}, where

$$A_{jk} = \langle\phi_{j0}|\hat{\boldsymbol{\varepsilon}}.\boldsymbol{P}\exp(\pm i\boldsymbol{k}.\boldsymbol{r})|\phi_{k0}\rangle \equiv \int \phi_{j0}^*(\hat{\boldsymbol{\varepsilon}}.\boldsymbol{P}\exp(\pm i\boldsymbol{k}.\boldsymbol{r}))\phi_{k0}\,dv \tag{2.106}$$

and, for convenience, the Dirac notation is now used. When the wavelength of the radiation, λ, is much larger than the atomic dimensions, the product $\boldsymbol{k}.\boldsymbol{r}$ will be much smaller than one. In these circumstances the expansion

$$\exp(i\boldsymbol{k}.\boldsymbol{r}) = 1 + i\boldsymbol{k}.\boldsymbol{r} - (1/2)(\boldsymbol{k}.\boldsymbol{r})^2 \tag{2.107}$$

can be approximated to unity. This is called the dipole approximation because the values of A_{jk} in (2.106) are now proportional to the matrix elements of the electric dipole moment. To show this, the approximation is used in the equation, to give

$$A_{jk} = \langle \phi_{j0} | \hat{\boldsymbol{\varepsilon}} . \boldsymbol{P} | \phi_{k0} \rangle \tag{2.108}$$

or equivalently

$$A_{jk} = m \langle \phi_{j0} | \hat{\boldsymbol{\varepsilon}} . \dot{\boldsymbol{R}} | \phi_{k0} \rangle \tag{2.109}$$

where m is the electron mass. Now, the time differential in this last equation can be removed using the Heisenberg equation of motion for $\boldsymbol{R}$. Thus,

$$\frac{d\boldsymbol{R}}{dt} = \frac{i}{\hbar}(H\boldsymbol{R} - \boldsymbol{R}H) \tag{2.110}$$

where the right hand side is the commutator operator of the Hamiltonian, H, and the vector $\boldsymbol{R}$. Inserting this equation into (2.109) gives

$$A_{jk} = \frac{im}{\hbar}(E_j - E_k) \langle \phi_{j0} | \hat{\boldsymbol{\varepsilon}} . \boldsymbol{R} | \phi_{k0} \rangle \tag{2.111}$$

where $\boldsymbol{R}$ represents the vector sum

$$\boldsymbol{R} = \sum_j \boldsymbol{r}_j \tag{2.112}$$

Since $\hat{\boldsymbol{\varepsilon}}$ is in the direction of the electric field vector (Equation (2.95)) the elements A_{jk} are proportional to the matrix elements of the atomic dipole moment. A further simplification is possible because of the well defined parity of any atomic state. Because $\boldsymbol{R}$ is an odd function the state $|\phi_k\rangle$ must have opposite parity to the state $|\phi_j\rangle$, and it therefore follows that the matrix A has no diagonal elements. The coupled equations can now be written

$$\dot{C}_2(t) = iC_1(t) d_{21}(\exp(i(\omega + \omega_{12})t) + \exp(-i(\omega - \omega_{12})t)) \tag{2.113}$$

and

$$\dot{C}_1(t) = iC_2(t) d_{12}(\exp(i(\omega - \omega_{12})t) + \exp(-i(\omega + \omega_{12})t)) \tag{2.114}$$

d_{21} and d_{12} are given by the equations

$$d_{21} = \frac{iq}{2\hbar} A_0 \omega_{12} \langle \phi_{20} | \hat{\boldsymbol{\varepsilon}} . \boldsymbol{R} | \phi_{10} \rangle \tag{2.115}$$

and

$$d_{12} = \frac{-iq}{2\hbar} A_0 \omega_{12} \langle \phi_{20} | \hat{\boldsymbol{\varepsilon}} . \boldsymbol{R} | \phi_{20} \rangle \tag{2.116}$$

If the probability of atomic excitation is small, that is for times such that C_2 is much less than C_1, C_1 can be made constant. C_2 is now given by Equation (2.113) with $C_1(0) = 1$. Integrating this directly gives

$$iC_2(t) = d_{21} \left(\frac{\exp(i(\omega + \omega_{12})t) - 1}{i(\omega + \omega_{12})} + \frac{\exp(-i(\omega - \omega_{12})t) - 1}{-i(\omega - \omega_{12})} \right) \tag{2.117}$$

Since the concern is for processes at or near resonance ($\omega \simeq \omega_{12}$) the first term in this equation can be neglected. This is known as the rotating wave approximation (RWA). In these circumstances, the probability of the atom being in state 2 is given by the expression

$$P_2(t) = |C_2(t)|^2 = \frac{4(d_{21})^2 \sin^2[(\omega - \omega_{12})t/2]}{(\omega - \omega_{12})^2} \tag{2.118}$$

This is the same as the function (2.46) with $\omega_{sn} = \omega - \omega_{12}$. (It is plotted in Figure 2.4.) For monochromatic radiation, the probability is a quadratic function of time at resonance ($\omega = \omega_{12}$), and oscillates with time away from resonance. Values for $P(t)$ in these circumstances are similar to the expressions (2.47) and (2.48), given previously. It is important to realize that these equations refer to an excited state, which has been approximated to a well defined stationary state, and whose time dependence is given by Equation (2.29). They therefore neglect any radiative broadening, and in this sense, Equation (2.118) is an artifact since it refers to an excited level with an infinite lifetime. However, this equation can still be used to calculate transition rates if the incident beam has a broad spectral distribution. To do this, we first relate the energy density to the amplitude of the wave at frequency ω. Thus the average amplitude A_0, for frequencies between ω and $\omega + d\omega$ is

$$W(\omega)\, d\omega = \tfrac{1}{2} \varepsilon_0 \omega^2 A_0^2 \tag{2.119}$$

Note carefully that this is different to (2.97) because we are now using an energy per unit frequency interval. Thus, the transition probability for frequencies between ω and $\omega + d\omega$ is obtained by first relating $|d_{21}|^2$ to $W(\omega)$ using (2.119) and (2.115), and then substituting into (2.118). The total probability for all frequencies in the interval from $\omega - \Delta\omega$ to $\omega + \Delta\omega$ is

$$P(t) = \frac{2q^2}{\varepsilon_0 \hbar^2} |\langle \phi_{20} | \boldsymbol{\varepsilon} . \boldsymbol{R} | \phi_{10} \rangle|^2 \int_{\omega - \Delta\omega}^{\omega + \omega\Delta\omega} W(\omega) \frac{\sin^2((\omega - \omega_{12})t/2}{(\omega - \omega_{12})^2} d\omega \tag{2.120}$$

which can be integrated directly when $W(\omega)$ is constant over the interval $2\Delta\omega$. The evaluation of this integral was discussed in Section 2.4. When $\Delta\omega t$ is much less than one it is proportional to t^2 whilst for $\Delta\omega t$ greater than one† the integral is

$$\int_{\omega-\Delta\omega}^{\omega+\Delta\omega} \frac{\sin^2((\omega-\omega_{12})t/2)}{(\omega-\omega_{12})^2}\,d\omega = \frac{\pi t}{2} \tag{2.121}$$

The total probability is therefore

$$P_2(t) = W(\omega)\frac{\pi q^2}{e_0\hbar^2}|\langle\phi_{20}|\boldsymbol{\varepsilon}(\omega)\,.\,\boldsymbol{R}|\phi_{10}\rangle|^2 t \tag{2.122}$$

where $\omega = \omega_{12}$ has been substituted to abbreviate the notation. If the concern is for randomly polarized radiation, this equation has to be averaged over all the directions, defined by $\hat{\boldsymbol{\varepsilon}}$. Since the average value of $\cos^2\theta$ is $1/3$, the probability is

$$P_2(t) = W(\omega)\frac{\pi q^2}{\varepsilon_0\hbar^2}\times\frac{1}{3}|\langle\phi_{20}|\boldsymbol{R}|\phi_{10}\rangle|^2 t \tag{2.123}$$

Of course, Equation (2.123) also applies for polarized radiation if the atoms are not aligned.

Now consider the same physical situation evaluated in terms of the Einstein coefficients. If the number of atoms in the upper state is N_2 and the density in the lower state is N_1, then

$$N_2 = N_0|C_2|^2 \tag{2.124}$$

and

$$N_1 = N_0|C_1|^2 \tag{2.125}$$

where $N_0 = N_1 + N_2$ is the total number of atoms. (Note that in this case only two levels are involved.) According to the definitions of A and B, the rate equation for the number N_2 is

$$\frac{dN_2}{dt} = -N_2A + WB(N_1 - N_2) \tag{2.126}$$

which can be written

$$\frac{dN_2}{dt} = -N_2A + WB(N_0 - 2N_2) \tag{2.127}$$

† This is for the frequency interval $\Delta\omega$ much larger than the natural broadening and for times comparable or greater than the natural lifetime, as we shall see later.

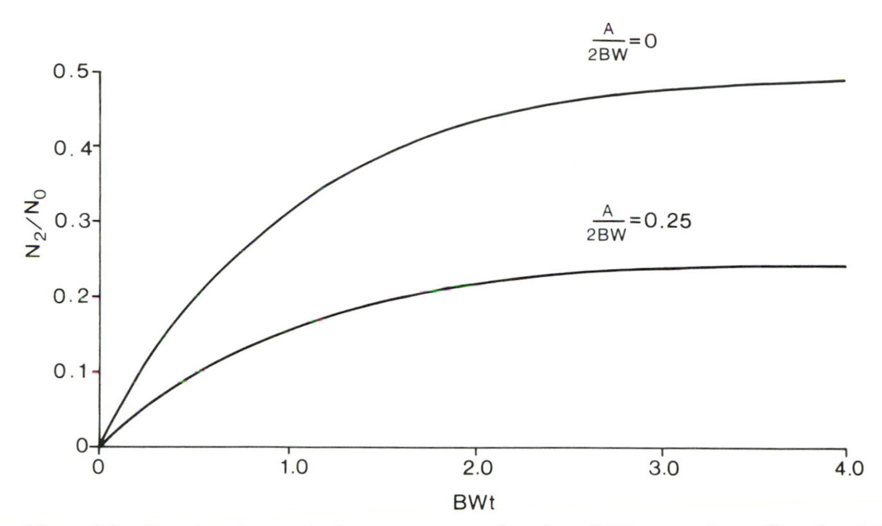

Figure 2.6. Fraction of atoms in the upper state as a function of BWt, according to Equation (2.128).

W is the energy density in the wave and, for simplicity, we have assumed that the levels are non-degenerate. Equation (2.125) can be integrated directly to give

$$N_2 = \frac{N_0 BW}{A+2BW}(1-\exp(-At-2BWt)) \tag{2.128}$$

where t is the time elapsed from switching on the beam. This is plotted in Figure 2.6. Now for small times, where $N_2 \ll N_0$, this approximates to

$$N_2 = N_0 BWt \tag{2.129}$$

and therefore

$$P_2(t) = \frac{N_2}{N_0} = BWt \tag{2.130}$$

For very long times the number of atoms in the upper state approaches the limit

$$N_2 = \frac{N_0 BW}{A+2BW} \tag{2.131}$$

Equation (2.131) shows that N_2 approaches the value $N/2$ when $2BW$ is much greater than A. This criterion can often be achieved using laser beams and when this happens the transition is said to be saturated. Any increase in laser power produces a negligible charge in the excited level population. The equation for small t (2.130) has the same form as (2.123), and therefore

$$B = B_{12} = \frac{\pi q^2}{3\varepsilon_0 \hbar^2} |\langle \phi_{20} | \boldsymbol{R} | \phi_{10} \rangle|^2 \tag{2.132}$$

By this means, the Einstein absorption coefficient has been related to fundamental atomic properties. Values for A_{21} and B_{21} can now be obtained using the Equations (2.24) and (2.25).

2.8. Coupled equations with radiative damping

The previous analysis refers to broad-band illumination, that is where the linewidth of the atomic transition is smaller than the spectral distribution of the incident radiation. Also, it assumes that the probability of an individual atom in its state 1 absorbing a photon is independent of the intensity of the beam. However, if an intense narrow-band laser beam is made to interact with atoms, the number of atoms in the excited state can often oscillate before a steady state is reached. These are known as Rabi oscillations, and this behaviour is in contrast to the smooth time variation predicted by the Einstein theory (Equation 2.126). In order to understand this, the coupled equations (2.111) and (2.112) are used, but now more exact solutions are sought. The analysis can be done directly using these two equations or it can include the effects of the finite broadening of the transitions. In the former case the oscillations are undamped, whilst in the latter they decrease in amplitude dependent on the transition broadening. This broadening can be included in the equations in a phenomenological way by adding an extra term to the two equations. Thus, in the rotating wave approximation

$$\dot{C}_2 = (iC_1\Omega_{21}/2)\exp(-i\omega t + i\omega_{12} t) - \gamma C_2 \tag{2.133}$$

and

$$\dot{C}_1 = (iC_2\Omega_{12}/2)\exp(i\omega t - i\omega_{12} t) \tag{2.134}$$

Here it is assumed that the upper level decays with a lifetime determined by γ. For an isolated two-level system Equation (2.134) should also be modified to account for repopulation. This cannot be done in a simple way and so the equations only strictly apply for the case when the upper level decays to levels unconnected to the lower level. If, however, we also include the unitarity condition (2.31) then equations can be constructed which are essentially the same as a more rigorous derivation. (The repopulation is thus introduced via Equation (2.31).) The two equations (2.133) and (2.134) have now been written in terms of the matrix elements Ω_{21} and Ω_{12}, which for plane polarized light, can be written

$$\Omega_{21} = -i\boldsymbol{E}_0 . \boldsymbol{D}/\hbar \qquad (2.135)$$

and

$$\Omega_{12} = i\boldsymbol{E}_0 . \boldsymbol{D}/\hbar \qquad (2.136)$$

where $\boldsymbol{D}$ is the dipole moment of the atom and is given by the expression

$$\boldsymbol{D} = q\langle\phi_{10}|\boldsymbol{R}|\phi_{20}\rangle = q\langle\phi_{20}|\boldsymbol{R}|\phi_{10}\rangle \qquad (2.137)$$

$\boldsymbol{E}_0$ is the electric field vector representing the amplitude of the incident wave and $\boldsymbol{R}$ is defined previously by (2.112). Notice that Ω_{12} and Ω_{21} have the dimensions of angular frequency because $\boldsymbol{E}_0 . \boldsymbol{D}$ is the energy of the dipole. Also, the complex nature of the matrix elements is arbitrary. We could have equally started with the vector potential $-\boldsymbol{A}_0 \sin \omega t$ (equivalent to $\boldsymbol{E}_0 \cos \omega t$), in which case Ω_{21} and Ω_{12} would both be real. Evidently, general relations for any phase are

$$\Omega_{21} = \Omega_{12}^* \qquad (2.138)$$

and

$$|\Omega_{21}|^2 = |\Omega_{12}|^2 = \Omega^2 \qquad (2.139)$$

In the absence of any oscillating electric fields, Equation (2.133) can be integrated directly to give

$$C_2(t) = C_2(0)\exp(-\gamma t) \qquad (2.140)$$

and hence the number of atoms in the upper state is related to the number at time zero, by the equation

$$N_2(t) = N_2(0)\exp(-2\gamma t) = N_2(0)\exp(-t/\tau) \qquad (2.141)$$

τ is now the lifetime of the excited state. Since this must be equivalent to (2.127), when $W = 0$, then

$$\tau = \frac{1}{2\gamma} = \frac{1}{A_{21}} \qquad (2.142)$$

If the lower level also decays by spontaneous emission (or some other mechanism) then a damping term also needs to be added to (2.134) in the same manner as (2.133). When there is no repopulation then the two equations can be solved

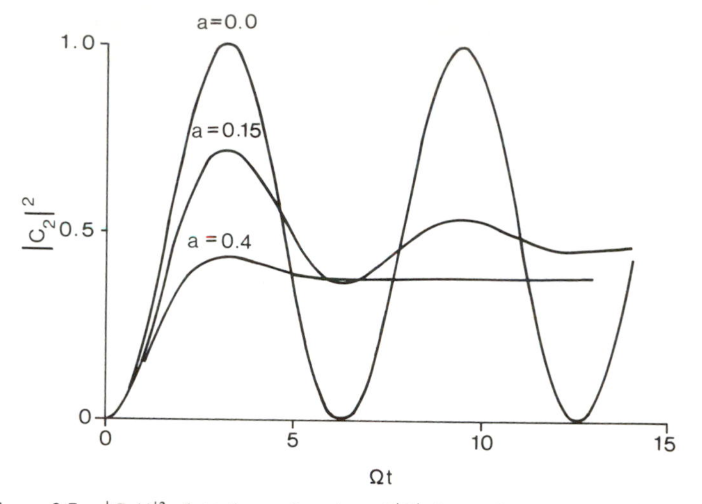

Figure 2.7. $|C_2(t)|^2$ plotted as a function of $|\Omega|t$ for various values of the ratio $a=\gamma/|\Omega|$.

directly, but notice that (2.31) does not apply since the number of atoms will decay exponentially.

For the closed two level system considered here a set of equations can be constructed from (2.133), (2.134) and (2.31) which involve only the terms $|C_1|^2$, $|C_2|^2$, $C_1C_2^*$, $C_2C_1^*$ and their time derivatives. These are known as the optical Bloch equations and all the measurable quantities, such as absorption coefficients, can be related to their solutions. The general details of these equations and their method of solution†, with or without damping, are quite complicated and will not be considered here. As an example of the type of solution which is obtained, we consider the case where $\omega = \omega_{12}$ and radiation damping is included. The probability of the atom being in the upper level is then

$$P_2(t) = |C_2(t)|^2 = \frac{\Omega^2}{2(2\gamma^2+\Omega^2)}\left(1-\left(\cos\lambda t+\frac{3\gamma}{2\lambda}\sin\lambda t\right)\exp(-3\gamma t/2)\right) \tag{2.143}$$

where

$$\lambda = \left(\Omega^2-\frac{\gamma^2}{4}\right)^{1/2} \tag{2.144}$$

Values of $|C_2|^2$ are shown in Figure 2.7 for different values of the radiative damping γ. When γ is zero the probability of the atom being in state 2 oscillates between 0 and 1, at an angular frequency of $\Omega/2$. Ω is known as the Rabi frequency. This effect is known as optical nutation and can be pictured in the

† Analytical solutions are not possible in all cases.

following way. If an intense monochromatic light beam is suddenly applied to atoms, the number of atoms in the excited state rapidly increases, providing the resonance condition is satisfied. This rate of increase is large and for a short time, the number in the excited state exceeds the number in the ground state. At this point stimulated emission starts to operate and the number starts to decrease. (The oscillations are in fact a manifestation of the dynamic Stark effect which is important because of the extremely high electric fields associated with the beam.) Evidently, the effect of spontaneous emission is to damp out the oscillations between resonance absorption and stimulated emission. For zero damping, which of course is physically unrealistic, the average value is still 0·5, in line with Equation (2.131). As the value of γ is increased the oscillations disappear, and the curves become similar to those predicted using the Einstein coefficients (Figure 2.6). Of course, the rate equation (2.126) refers to broad-band illumination but the general equivalence between the two can be established by integrating over the line profile. A similar analysis was used to relate B_{12} to the atomic matrix elements but here stimulated emission was playing no part.

The coupled or Bloch equations therefore give a 'reasonable explanation' for stimulated emission, that is they explain the phase, direction and intensity of the emitted wave. (They are analogous to the classical situation of coupled oscillators.) Now it is obvious that this interpretation has limitations. The C_2 coefficients can only be related to the number of atoms in the excited state (Equation (2.122)) by averaging over a larger number of atoms. As soon as one asks about individual atoms problems arise which can only be adequately reconciled with a properly quantized radiation field.

2.9. *Widths and profiles of spectral lines*

If an atom decays spontaneously the emitted radiation is not completely monochromatic, but has a smooth distribution of frequencies centred about ω_{12}, determined by Equation (2.18). If the lifetime of the excited state is τ, then the uncertainty principle demands an average spread in angular frequency of the order $1/\tau$ ($\Delta E \,.\, \tau \simeq \hbar$). This is the radiative damping mentioned in the previous section. Similarly, if radiation is absorbed from an external source, the absorption occurs across a range of frequencies. The profile and width of the distribution in both cases is not, however, determined by the natural lifetime alone. There are several other broadening effects to be considered, and only in the limit of isolated atoms at very low temperatures will the line width approach the value given by the radiative value. The other principle contributions, considered here, are Doppler, pressure and power broadening. Doppler broadening is an example of an inhomogeneous broadening mechanism because the absorption is altered for different groups of atoms in the ensemble. It is associated with a Gaussian line shape. When the Doppler effects are small the line is homogeneously broadened, in the absence of electric and magnetic fields, and has a Lorenzian profile.

Consider first the radiative or natural linewidth. Equation (2.133) can be solved directly for the case of a weak incident beam. For algebraic simplicity we retain the RWA implicit in these equations. The first order approximation is obtained by putting $C_1 = 1$ into Equation (2.131) and solving directly. A solution with the correct boundary condition, $C_2(0) = 0$, is

$$C_2(t) = \frac{\Omega_{21}(\exp(-i\omega t + i\omega_{12}t) - \exp(-\gamma t))}{2((\omega_{12} - \omega) - i\gamma)} \tag{2.145}$$

and therefore

$$\begin{aligned} P_2(t) &= |C_2(t)|^2 \\ &= \frac{\Omega^2}{4((\omega_{12} - \omega)^2 + \gamma^2)}(1 - 2\exp(-\gamma t)\cos(\omega_{12}t - \omega t) + \exp(-2\gamma t)) \end{aligned} \tag{2.146}$$

This equation will only be valid if $\Omega \ll \gamma$. For broad-band illumination, it can be integrated with respect to frequency. The solution corresponds to the rate equation (2.128) when $BW \ll A$, with the Einstein coefficients derived from (2.132), (2.24) and (2.25). The constant multiplier in the equation has a Lorenzian frequency profile with a linewidth 2γ, equal to the inverse of the natural lifetime. Most often, the quantity measured in experiments is the absorption coefficient K, which is related to the change in intensity of a beam traversing an optical sample, by the equation

$$I = I_0 \exp(-Kz) \tag{2.147}$$

Here z is the distance along the direction of propogation and I is the intensity. This equation only applies for a weak beam, for very intense laser beams the variation is linear rather than exponential. Now K can be related to $|C_2|^2$ in the following manner. The change in intensity traversing a thin section can be written

$$\delta I = -IK\,\delta z \tag{2.148}$$

and this can be related to the energy density, W, by considering the energy balance in this section. Thus

$$\delta I = \frac{\partial W}{\partial t}\delta z \tag{2.149}$$

and K is therefore

$$K = -\frac{1}{I}\frac{\partial W}{\partial t} \tag{2.150}$$

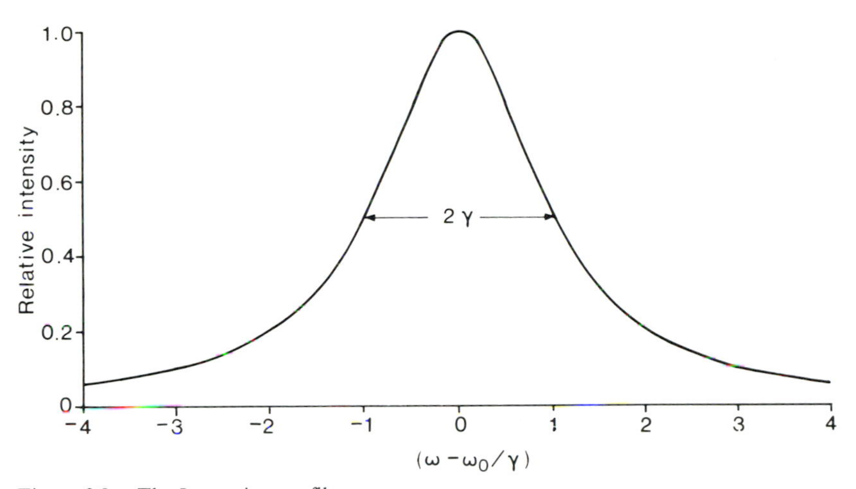

Figure 2.8. The Lorenzian profile.

In the 'steady state' condition the derivative on the right hand side can be equated to the radiative energy loss of the excited atoms. If there are $n_2(\omega)$ atoms per unit volume, this is just $(-n_2(\omega)\hbar\omega/\tau)$, and the atomic absorption coefficient is

$$K = \frac{n_2(\omega)\hbar\omega}{I_\omega \tau} \tag{2.151}$$

Relating $n_2(\omega)$ to the probability (2.143), using Equation (2.122), and substituting this into (2.146) gives

$$K(\omega) = \frac{\Omega^2 \gamma}{2I_\omega((\omega_{12}-\omega)^2+\gamma)} \tag{2.152}$$

for the absorption at times much greater than τ. Finally, the expressions for Ω_{21} and I_ω as given by (2.135) and (2.99) (now modified to account for the dielectric constant), can be substituted into this equation to give†

$$K(\omega) = \frac{1}{3}\frac{n_0 q^2}{\eta(\omega)\hbar\varepsilon_0 c}\frac{\omega_{12}\gamma|R|^2}{((\omega_{12}-\omega)^2+\gamma^2)} \tag{2.153}$$

where the abbreviation $|R|^2 = |\langle\phi_2|\boldsymbol{R}|\phi_1\rangle|^2$ has been used. η is the refractive index which can be approximated to unity and ω has been replaced by ω_{12} on the top line because the RWA has been used; the interest being in frequencies relatively close to the line centre. The frequency dependence of the absorption coefficient has a Lorenzian shape with a full width at half height of 2γ (or $1/\tau$). This is plotted in Figure 2.8.

† Note the factor 1/3 to account for random orientation of the field and dipole moment.

It may be thought from the previous analysis that the coupled equations only contain the information to calculate absorption, and dispersion is neglected. In fact they can be used to calculate the macroscopic properties of a gas which are relevant to the propagation of a light beam. Consider the wave equation for the vector potential, in a medium with a dielectric constant ε. This is written

$$\nabla^2 \boldsymbol{A} - \varepsilon\varepsilon_0 \boldsymbol{A} = 0 \tag{2.154}$$

A plane wave solution propogating in the z direction has the form

$$\boldsymbol{A} = \boldsymbol{A}_0 \exp(-\omega\kappa z/c)\exp(i\omega t - i\omega\eta z/c) \tag{2.155}$$

where η is the refractive index. Evidently η and κ are related by the equation

$$(\eta - i\kappa)^2 = \varepsilon = 1 + \chi \tag{2.156}$$

where χ is the susceptibility. The amplitude of the wave is therefore attenuated exponentially according to Equation (2.155). Since the intensity depends on $|\boldsymbol{A}|^2$ (Equation 2.99) the absorption coefficient is just

$$K = 2\omega\kappa/c \tag{2.157}$$

Using (2.156) this can be written

$$K = (\omega/\eta c)\mathrm{Im}(\chi) \simeq (\omega/c)\mathrm{Im}(\chi) \tag{2.158}$$

Similarly, the refractive index is given by the equation

$$\eta = (1 + \mathrm{Re}(\chi) + \kappa^2)^{1/2} \simeq 1 + \mathrm{Re}(\chi)/2 \tag{2.159}$$

Here, the real and imaginary parts are abbreviated by Re and Im respectively. Both the absorption and dispersion can therefore be calculated from the susceptibility. If an oscillating field $\boldsymbol{E}\sin\omega t$ is applied to a dielectric, then the susceptibility is defined by the relation

$$\boldsymbol{P}\sin(\omega t) = \varepsilon_0\chi(\omega)\boldsymbol{E}\sin\omega t \tag{2.160}$$

$\boldsymbol{P}$ is the macroscopic polarization of the dielectric and χ can have complex components to account for any resulting phase change. It is convenient to consider the electric field is along the x direction, in which case the polarization is related to the x component of the microscopic dipole moment d and the number density of atoms n_0, by

$$P = n_0 d \tag{2.161}$$

d is defined by the equation

$$d = qX \tag{2.162}$$

where X is the x component of the vector $\boldsymbol{R}$ used previously. In the electric field, represented by the vector potential (2.101), the dipole oscillates and we need to calculate its explicit time dependence. This is contained in the equation

$$d(\varepsilon) = q\langle\psi(t)|X|\psi(t)\rangle \tag{2.163}$$

The total wave function ψ, as given by (2.30) and (2.29), is then substituted into this equation to give

$$d(t) = q(C_1^* C_2\langle\phi_{10}|X|\phi_{20}\rangle \exp(-i\omega_{12}t) + C_2^* C_1\langle\phi_{20}|X|\phi_{10}\rangle \exp(i\omega_{12}t)) \tag{2.164}$$

For the weak field approximation, $C_1 = 1$ and C_2 is given by the oscillating part of (2.145). (The exponentially decaying part has no consequence here.) Substituting this value for C_2 into (2.164), and using (2.135) and (2.136), results in the expression

$$d(t) = \frac{\hbar\Omega^2}{((\omega_{12}-\omega)-i\gamma)}\frac{(-\exp(-i\omega t)+\exp(i\omega t))}{2iE_0} = \frac{\hbar\Omega^2}{((\omega_{12}-\omega)-i\gamma)}\frac{\sin\omega t}{E_0} \tag{2.165}$$

Now the applied field, as represented by the vector potential (2.101), has a time dependence $E_0 \sin\omega t$, so the susceptibility is just

$$\chi(\omega) = \frac{n_0\hbar\Omega^2}{\varepsilon_0((\omega_{12}-\omega)-i\gamma)} \tag{2.166}$$

If the imaginary part of this is now substituted into (2.158), and averaged over all orientations of $\boldsymbol{E}$ and $\boldsymbol{R}$, the formula for $K(\omega)$, as given by Equation (2.152), is obtained. The refractive index is found by putting the real part of $\chi(\omega)$ into Equation (2.158). This can then be written

$$\eta(\omega) = 1 + \frac{1}{2}\frac{(\omega_{12}-\omega)}{\gamma}\left(\frac{c}{\omega_{12}}\right)K(\omega) \tag{2.167}$$

where $K(\omega)$ is the absorption coefficient. Both K and η are plotted in Figure 2.9. Near to resonance, when $\omega = \omega_{12}$, the refractive index undergoes a rapid change which results in a phase change of the propagating beam. This is caused by the phase lag between the induced dipole oscillation and the radiation field.

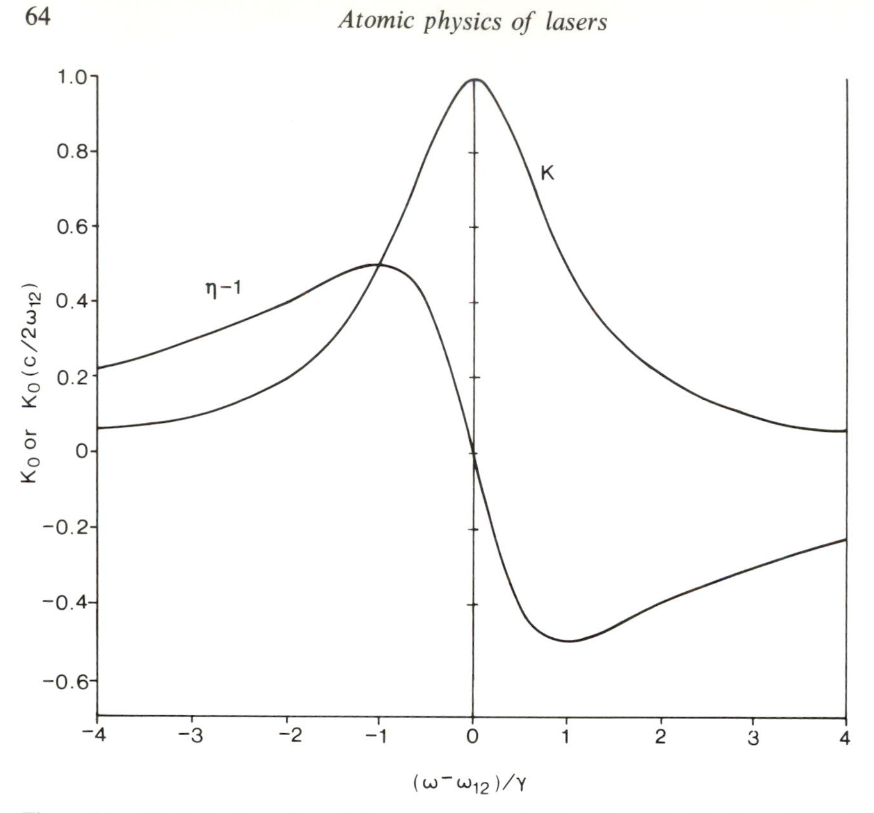

Figure 2.9. The behaviour of η and K near resonance. K is in units of K_0, the absorption coefficient when $\omega = \omega_{12}$, and $\eta - 1$ is in units of $K_0(c/2\omega_{12})$.

It is convenient now to consider how the lineshape changes as the beam intensity is increased. For this purpose, more general solutions to the Bloch equations are sought, rather than the restricted one given by (2.146). Explicit expressions for $|C_1|^2$ and $|C_2|^2$ can be found for any ω, when $t \gg \tau$. The steady state solution for $|C_2|^2$, which has the same frequency profile as the absorption coefficient, is

$$|C_2|^2 = \frac{\Omega^2}{4\left((\omega_{12}-\omega)^2 + \dfrac{\Omega^2}{2} + \gamma_n^2\right)} \tag{2.168}$$

When Ω is much greater than γ_n (or the detuning $\omega - \omega_0$), $|C_2|^2$ is 0·5, and the transition is saturated. Because Ω^2 is proportional to $|E_0|^2$ the increase in the linewidth depends on the beam intensity and is known as power broadening (or saturation broadening). The total linewidth, $2\gamma_t$, is

$$2\gamma_t = 2\left(\frac{\Omega^2}{2} + \gamma_n^2\right)^{1/2} \tag{2.169}$$

where $2\gamma_n$ is the natural linewidth. The absorption coefficient can be calculated in a similar manner to previously, giving the expression (2.153), but with the denominator adjusted to include the additional term $\Omega^2/2$.

Up to now, the assumption has been made that the atoms are isolated from another, and their only interaction is with the radiation. This is reasonable for a low pressure gas when the time between atomic collisions is much larger than the natural lifetime. Each atom absorbs and re-emits photons many times between collisions so any average property such as the absorption coefficient is unaltered. When the time between collisions becomes comparable to τ this no longer applies. The collisions can be divided into two types, inelastic scattering where the atom de-excites, and elastic scattering where the excited atom remains in the higher atomic level. The effect of the first process, collision induced transitions, is to reduce the lifetime of the atomic state in accordance with the equation

$$\frac{1}{\tau} = \frac{1}{\tau_n} + \frac{1}{t_c} \tag{2.170}$$

where τ_n is the natural lifetime and t_c is the mean time between collisions which de-excite the atom. If this new value of τ is used in the optical Bloch equations, the steady state solution for the population of the excited states is

$$|C_2|^2 = \frac{\Omega^2}{4\left((\omega_{12} - \omega)^2 + \dfrac{\gamma_t}{\gamma_n}\Omega^2 + \gamma_t^2\right)} \tag{2.171}$$

where

$$\gamma_n = \frac{1}{2\tau_n} \tag{2.172}$$

and

$$\gamma_t = \gamma_n + \frac{1}{2t_c} = \gamma_n + \gamma_c \tag{2.173}$$

Notice that the absorption coefficient (2.153) will now have γ_t in the numerator, as well as the denominator changing in line with Equation (2.170). When no power broadening is present the linewidth is increased from $2\gamma_n$ to $(2\gamma_n + 2\gamma_c)$. According to the kinetic theory of gases the mean time between collisions of any type is

$$\bar{t} = \left(\frac{m}{\pi k T}\right)^{1/2} \frac{1}{16 R_0^2 n_0} \tag{2.174}$$

In this equation R_0 is the atomic 'collision' radius and m is the mass of the atom. Now the relation between t_c and $\bar{t}$ is not simple because the collisions which induce transitions are, among other things, velocity dependent. However, the width γ_c, associated with collision induced transitions can be expressed in parametric form, as

$$\gamma_c = f_1(pT)/\bar{t} = f_2(pT)p \tag{2.175}$$

where f_1 and f_2 are dimensionless constants depending on the pressure and temperature T. f_1 may be thought of as the fraction of collisions which induce the excited atom to decay. It turns out that this type of broadening is often much smaller than that caused by elastic collisions, a fact which indicates that the majority of collisions are elastic. (This of course will depend on the atom or molecule in question, the nature of the states involved and the pressure and temperature!) For elastic collisions the broadening is brought about by interruptions of the phase between the atoms and the interacting light beam. A detailed analysis is quite complicated and will not be carried out here. If the collision excitations are ignored then the broadening due to this effect can be approximated by the equation

$$\gamma_\phi = 1/\bar{t} \tag{2.176}$$

When no power broadening is present the total width can be obtained from the relation

$$\gamma_t = \gamma_n + (2A_\phi + A_c)p/2 \tag{2.177}$$

where p is the pressure. This is commonly called pressure broadening. A_ϕ and A_c are the pressure broadening 'constants' which refer to elastic and inelastic scattering, respectively. Both A_ϕ and A_c depend on pressure and temperature, but when A_c is small, A_ϕ becomes approximately pressure independent, and is found from the two equations (2.174) and (2.176).

Lastly, the broadening due to the Doppler effect is considered. This is a case of inhomogeneous broadening and is best understood in terms of the number density of atoms which contribute to the absorption coefficient. Roughly speaking, atoms will only absorb radiation if their velocity is such that the Doppler shifted frequency is within a few homogeneous line widths of the line centre, ω_{12}. The frequency of light in the rest frame of the atom is, to first order, $\omega_0 + \boldsymbol{k}.\boldsymbol{v}$, where $\boldsymbol{v}$ is the velocity and $\boldsymbol{k}$ is the wavevector. Consider, in the first instance, that the collision broadening can be neglected, and the atoms are interacting with a plane wave travelling in the z direction. The number of atoms with a z component of velocity between v and $v+dv$ is given by the Maxwell distribution

$$n(v)\,dv = n_0\left(\frac{m}{2\pi k_b T}\right)^{1/2} \exp(-mv^2/2k_b T) \tag{2.178}$$

where n_0 is the number density of atoms, m is the mass of the atom and k_b is the Boltzmann constant. (Not to be confused with the wavevector $\boldsymbol{k}$!). For each of these atoms the light is shifted from ω, the incident wave frequency, to $\omega(1-v/c)$. The susceptibility for this group is then obtained by substituting this value of frequency into (2.166) with 'n_0' replaced by Equation (2.178). The total susceptibility in the weak field limit, is therefore given by the integral

$$\chi(\omega) = \frac{n_0 \hbar \Omega^2}{\varepsilon_0 (\pi v_0)^{1/2}} \int_{-\infty}^{+\infty} \frac{\exp(-v^2/v_0^2)}{\omega_{12} - \omega(1-v/c) - i\gamma} \, dv \tag{2.179}$$

where v_0 is defined by

$$v_0^2 = 2k_b T/m \tag{2.180}$$

Now the absorption coefficient is obtained from the imaginary part of $\chi(\omega)$ in the manner outlined previously. Thus

$$K(\omega) = \frac{n_0 \hbar \Omega^2 \omega_{12} \gamma_n}{\varepsilon_0 c (\pi v_0)^{1/2}} \int_{-\infty}^{+\infty} \frac{\exp(-v^2/v_0^2)}{(\omega_{12} - \omega(1-v/c))^2 + \gamma_n^2} \, dv \tag{2.181}$$

This equation can be extended to include power broadening by substituting γ_t, as given by Equation (2.169) in place of the γ_n which appears in the denominator of the integrand. It may also be reasonably extended to account for collisional effects using γ_t, defined by Equation (2.177). The result is only approximate for this case because the collisions are related to changes in the velocity, which affect the Doppler shifts. These can, in some circumstances, lead to a reduction in the width of the lineshape, known as Dicke narrowing.

The integral in (2.181) has no general analytical solution, but can be evaluated using numerical techniques. The resulting lineshape is known as a Voigt profile†. In the limiting case, when the homogeneous broadening is much smaller than the Doppler broadening, the absorption coefficient has the Gaussian lineshape

$$K(\omega) = K_0 \exp\left[\frac{(\omega_{12} - \omega)^2 c^2}{\omega_{12}^2 v_0^2}\right] \tag{2.182}$$

which has a full width at half height,

$$\Delta_I = 1{\cdot}666 \omega_{12} (v_0/c) \tag{2.183}$$

Figure 2.10 compares this lineshape with the Lorenzian. Even when the

† Tabulations can be found, for example, in Fried, B.D. and Conte, S.D., 1961, *The Plasma Dispersion Function* (New York: Academic Press).

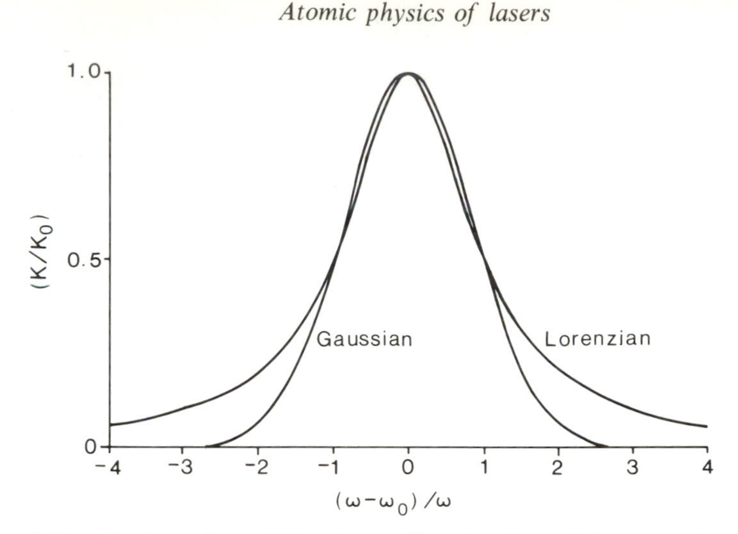

Figure 2.10. The Lorenzian and Gaussian profiles when their widths are equal.

homogeneous broadening is much smaller than the Doppler broadening, the tail of the distribution away from resonance is dominated by the Lorenzian. The composite lineshape, or Voigt profile, has a full width which is between the two limits

$$\Delta_c^2 = \Delta_I^2 + \Delta_H^2 \tag{2.184}$$

and

$$\Delta_c = \Delta_I + \Delta_H \tag{2.185}$$

Here, ΔH is given by the equation

$$\Delta H = (\gamma_t \Omega^2/\gamma_n + \gamma_t^2)^{1/2} \tag{2.186}$$

where γ_ω is given by (2.177). This last equation is only approximate since it includes collision broadening.

It is useful at this point to quote some average values for the widths of spectral lines. The lifetime associated with a strong transition in the visible part of the spectrum is of order 10^{-8} s, giving a width $2\gamma_n$ of 16 MHz in frequency units (not angular frequency, which is 100 MHz here!). The full width Doppler broadening in a gas at room temperature (20°C) for atomic mass 60, and wavelength 500 nm, is approximately 948 MHz. Pressure broadening constants A_ϕ and A_c (Equation 2.174) are found experimentally to be in the range 1–30 MHz/torr, depending on the kind of atom (or molecule) and the temperature. A value of 25 MHz for A_ϕ would correspond to a 'hard sphere' radius of $3{\cdot}5 \times 10^{-10}$ m. To calculate the power broadening, for a monochromatic source, the value of $\Omega^2/2$ from (2.135) is evaluated in terms of the A coefficients using Equations (2.132) and (2.24). This last equation is modified to account for the factor 2π described in the footnote on page 35. Also, the value of Ω^2 needs to be averaged to allow for the different orientations of the electric field and the atomic dipole moment. The final expression, which applies for an incident beam whose linewidth is less than γ_n, is

$$\frac{\Omega^2}{2} = \frac{\pi I c^2}{\hbar \omega_{12}^3} A = \frac{\pi I c^2}{\hbar \omega_{12}^3} 2\gamma_n \tag{2.187}$$

where, $I = \varepsilon_0 E_0^2 c/2$, has been used to relate the intensity, I, to the amplitude of the electric field, E_0. For the case considered above, the total linewidth give by expression (2.161) is doubled for an intensity of 0·8 watts/cm^2. Now, modern tunable lasers have linewidths often less than 1 MHz, and, using these sources, a strong visible transition will be almost completely saturated at intensities as low as 10 watts/cm^2. Consider what happens when such a monochromatic beam is incident on a low pressure gas. Consider, in the first instance, that collisional effects can be ignored. The number density of atoms (or molecules) in the excited state and the ground state as a function of atomic velocity are

$$n_2(v) = n_0(v)|C_2|^2 = \frac{n_0(v)\Omega^2}{4\left((\omega_{12} - \omega(1 - v/c))^2 + \dfrac{\Omega^2}{2} + \gamma_n^2\right)} \tag{2.188}$$

and

$$n_1(v) = n_0(v) - n_2(v) \tag{2.187}$$

where $n_0(v)$ has the Maxwell distribution (2.178), and represents the total number density of atoms in the gas. These apply for the simplified case where both states are isolated from relaxation effects. If the transition is in the infra-red (molecules), the excited state may be significantly populated by collisions in the gas. In this case, the equations can be simply modified by adding an extra term to (2.187) to represent the number of atoms in the absence of a light signal. (This will normally also have a Gaussian distribution of velocities.) Similarly the ground state may be significantly depopulated by relaxation to other levels but this can be accounted for by multiplying $n_0(v)$ by a constant fraction.

Equation (2.187) shows that the number n_2 increases for velocities close to the condition $\omega_{12} = \omega(1 - v/c)$. When the Doppler broadening is dominant, as it is in most cases, the distributions of atoms in both states looks like those in Figure 2.11. The number of atoms in the excited state forms a peak with an approximately Lorenzian shape and a total width $2\gamma_t(c/\omega)$. Correspondingly, a hole is made in the ground state velocity distribution, with the same width and a

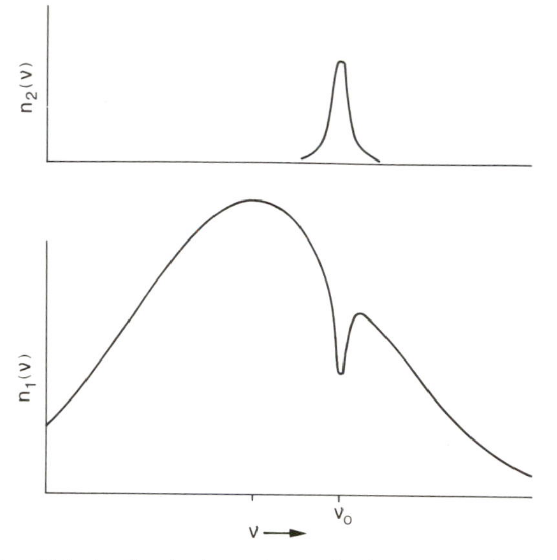

Figure 2.11. Velocity distributions of atoms in the presence of an intense monochromatic beam with frequency ν_0. The top graph is the excited state, whilst the lower one is the ground state. Velocity changing collisions are ignored.

depth which is proportional to $\Omega^2/(\gamma_n^2+\Omega^2/2)$. Even though a hole appears in the number of atoms the absorption coefficient is still symmetric, as determined by Equation (2.182), but now modified for power broadening. This just means that the hole would move across the velocity profile as the laser is scanned in frequency. However, it is possible to detect the presence of the hole using another laser beam. For example, a second weak laser beam would exhibit an absorption coefficient proportional to (n_2-n_1). (At saturation, $n_2-n_1 = 0$, and there will be no absorption.) If the first intense beam is reflected back into the gas along the opposite path, a strong dip will appear in the absorption spectrum at the centre of the Doppler profile. This is known as the Lamb dip and provides an accurate marker for the line centre ω_{12}; it forms the physical basis of saturated absorption spectroscopy.

When the gas pressure is increased the two distributions are modified by velocity changing collisions. Atoms in the excited state are removed from the narrow peak and spread across the Gaussian profile, whilst correspondingly, the hole depth is reduced. These are important considerations in the saturation characteristics of gas lasers and will be discussed more thoroughly in Chapter 4.

2.10. *Spontaneous and stimulated emission*

In Section 2.7 the absorption of radiation was calculated using the matrix elements of the electric dipole moment operator. Thus, the Einstein coefficient for absorption of this type of radiation is

$$B_{12} = \frac{\pi}{\varepsilon_0} \frac{q^2}{\hbar^2} \frac{1}{3} |\langle \phi_2 | \boldsymbol{R} |_1 \rangle|^2 \tag{2.190}$$

where ϕ_1 and ϕ_2 are the spatial wave functions† of the two states. The coefficient for B_{21} is simply obtained from Equation (2.25) and is

$$B_{21} = \frac{g_1}{g_2} \frac{\pi q^2}{\varepsilon_0 \hbar^2} \frac{1}{3} \langle \phi_2 | \boldsymbol{R} | \phi_1 \rangle^2 \tag{2.191}$$

Here g_1 and g_2 are the degeneracies of the levels E_1 and E_2. Now the spontaneous emission coefficient A_{21}, which is simply the inverse of the natural lifetime, can be obtained from Equation (2.24), taking care to express the energy density in terms of angular frequency. A_{21} is therefore given by the equation

$$A_{21} = \frac{g_1}{g_2} \frac{q^2 \omega_{12}^3}{\pi \varepsilon_0 \hbar c^3} \frac{1}{3} |\langle \phi_2 | \boldsymbol{R} | \phi_1 \rangle|^2 \tag{2.192}$$

These equations can be compared with a general expression for transition rates, calculated according to first order perturbation theory, using the QED Hamiltonian (2.102). Here the 'wavefunctions' for the initial and final states are not just the atomic wavefunctions, but include the quantized radiation field. If the whole system of an atom and the radiation field is contained in a 'box' of volume V, the probability per unit time, of the system changing from N_k photons to $N_k + 1$, is

$$\frac{\partial P_{2-1}}{\partial t} = \frac{\pi q^2 \omega_{12}}{\varepsilon_0 \hbar V} |\langle \phi_2 | \hat{\boldsymbol{\varepsilon}}_k . \boldsymbol{R} | \phi_1 \rangle|^2 (N_k + 1) \delta(\omega_{12} - \omega_k) \tag{2.193}$$

$\boldsymbol{k}$ is the wavevector of the photon and $\hat{\boldsymbol{\varepsilon}}_k$ is its polarization which according to Section 2.1 can have two values for each mode $\boldsymbol{k}$. This expression, of course, refers to dipole radiation for two non-degenerate states ϕ_1 and ϕ_2. It contains both stimulated emission and spontaneous emission. When N_k is zero, that is where no radiation is present initially, then $\partial P_2 / \partial t$ is still finite and refers to spontaneous emission. In this case the total probability is given by the sum of all the probabilities for each value of $\boldsymbol{\varepsilon}_k$. If the average value of the scalar product is taken it is appropriate to find the number of different values of $\hat{\boldsymbol{\varepsilon}}_k$ using the mode density in frequency space. The total probability per unit time is thus the sum over all the modes, which is

$$\frac{\partial P}{\partial t} = \sum_k \frac{\partial P_k}{\partial t} = \int \rho(\omega) \frac{\partial P_{2-1}(\omega)}{\partial t} \, d\omega \tag{2.194}$$

† These are ϕ_{10} and ϕ_{20} in the previous equations, for example, (2.132).

Using $\rho(\omega) = \omega^2/\pi^2 c^3$ and substituting (2.193) into (2.194) gives

$$A_{21} = \frac{\partial P}{\partial t} = \frac{q^2\omega_{12}}{3\pi\varepsilon_0 \hbar c^3} \langle \phi_2|\boldsymbol{R}|\phi_1\rangle^2 \int \omega^2 \delta(\omega_{12}-\omega)\, d\omega \tag{2.195}$$

where the factor of 1/3 comes from averaging the scalar product over all angles. This expression is identical to (2.192), derived previously.

The corresponding stimulated emission coefficient can also be calculated from (2.193) in a similar fashion. Here the photon goes into the same cavity mode so there is only one term in the summation. The expression can be adjusted for broadband illumination if the equation

$$W(\omega)\, d\omega = \frac{N\hbar\omega}{V} \tag{2.196}$$

is employed to relate the number of photons in the frequency interval from ω to $\omega + d\omega$, to the energy density, $W(\omega)$. The total transition probability is thus

$$\frac{\partial P}{\partial t} = \sum_{\omega} \frac{\partial P_{2-1}(\omega)}{\partial t} = \frac{\pi q^2 \omega_{12}}{\varepsilon_0 \hbar} |\langle \phi_2|\boldsymbol{\varepsilon}_k \,.\, \boldsymbol{R}|\theta_1\rangle|^2 \int \frac{w(\omega)}{\hbar\omega} \delta(\omega_{12}-\omega)\, d\omega \tag{2.197}$$

and this reduces to

$$\frac{\partial P}{\partial t} = B_{21} W(\omega_{12}) = \frac{\pi q^2 \omega_{12}}{3\varepsilon_0 \hbar^2} |\langle \phi_2|\boldsymbol{R}|\phi_1\rangle|^2 \tag{2.196}$$

which gives B_{21} identical to Equation (2.191). A similar QED expression to (2.193) can be written for the absorption of radiation (ϕ_1 and ϕ_2 are reversed!) and this can be used to derive the expression (2.190) for the B_{12} coefficient.

2.11. Higher order multipole terms

So far, the discussion has involved only electric dipole transitions. If the selection rules for this type of transition (dealt with in Section 2.12), are disobeyed then higher order terms in the expansion (2.107) are required. Each term is reduced approximately by the factor ($2\pi/\lambda \times$ atomic radius) from the preceding term, which is of the order of 10^{-3} for visible transitions. Generally speaking, higher order terms can be ignored, except where the electric dipole transitions are forbidden (or highly retarded), unless the wavelength is in the X-ray or vacuum ultraviolet region.

The series represented by (2.107) is not a convenient expansion. A more suitable form is in terms of the electric and magnetic multipoles. This expansion is the same as that used to calculate the electric and magnetic fields from a system of static electric charges and magnetic poles. Of course, these oscillate with the varying field, but the analysis in Section 2.7 showed how the time dependence could be removed from the calculation. ΔH can therefore be written as a series of terms representing the interaction between the electric and magnetic charge distributions and the amplitudes of the electric and magnetic fields. Omitting the electric monopole, because this cannot oscillate in an electromagnetic field, the terms of the series, in order of decreasing size, are, electric dipole, magnetic dipole, electric quadrupole, etc. A general expression for B_{12} can be written using a similar analysis to that used to derive Equation (2.132). Thus,

$$B_{12} = (2\pi/\hbar W(\omega))|\langle \phi_2|\overline{\Delta H}|\phi_1\rangle|^2 \tag{2.199}$$

where the line over ΔH indicates the average over all directions when this is appropriate. Now, since W and ΔH both depend quadratically on the electric or magnetic field amplitudes, the previous expression can be written as

$$B_{12}^{E} = (\pi/\varepsilon_0\hbar^2)|\langle\phi_2|\overline{\Delta\hat{H}}|\phi_1\rangle|^2 \tag{2.200}$$

for electric multipole radiation, and

$$B_{12}^{M} = (\pi\mu_0/\hbar^2)|\langle\phi_2|\overline{\Delta\hat{H}}|\phi_1\rangle|^2 \tag{2.201}$$

for magnetic multipole radiation. Here $\Delta\hat{H}$ is given in terms of the unit vectors of the electric or magnetic field. Expressions (2.200) and (2.201) are the semi-classical analogues of the QED Equation (2.193). From the previous analysis, the value of $\Delta\hat{H}$ for electric dipole radiation is

$$\Delta\hat{H}_{E1} = q\hat{\boldsymbol{\varepsilon}}.\boldsymbol{R} \tag{2.202}$$

where $\hat{\boldsymbol{\varepsilon}}$ is the unit electric field vector. The corresponding expression for magnetic dipole radiation is

$$\Delta\hat{H}_{M1} = \hat{\boldsymbol{\varepsilon}}_B.\boldsymbol{\mu}_L \tag{2.203}$$

Here $\boldsymbol{\mu}_L$ is the dipole moment associated with the orbital angular momentum of the electron. For many electrons this is given by the expression

$$\boldsymbol{\mu}_L = \frac{q}{2m}\sum_j \boldsymbol{p}_j \times \boldsymbol{r}_j = \frac{q}{2m}\sum_j \boldsymbol{l}_j \tag{2.204}$$

where $\boldsymbol{l}_j$ is the orbital angular momentum of the electron. $\Delta\hat{H}$ can be written

$$\Delta\hat{H}_{M1} = (q/2m)\hat{\boldsymbol{\varepsilon}}_B.\boldsymbol{L} \tag{2.205}$$

where $\boldsymbol{L}$ is the total orbital angular momentum. This expression does not include any relativistic effects since the starting point of the calculation is the classical Hamiltonian (2.88). In this context the most important correction is obtained by including the electron spin. Equation (2.205) is therefore modified to

$$\Delta\hat{H}_{M1} = \hat{\boldsymbol{\varepsilon}}_B \cdot \boldsymbol{\mu}_J \tag{2.206}$$

where $\boldsymbol{\mu}_J$ is the total dipole moment of the atom.

The next term in the multipole expansion involves the interaction between the charge distribution and the gradients of the electric fields. Obviously, this is most important when the wavelength is small, because only then will there be a significant variation across the dimensions of the atom. A general form for $\Delta\hat{H}_{E2}$ is

$$\Delta\hat{H}_{E2} = -\frac{1}{2}q\sum_j (\boldsymbol{r}_j \cdot \boldsymbol{\nabla})(\boldsymbol{r}_j \cdot \hat{\boldsymbol{\varepsilon}}(0)) \tag{2.207}$$

where the summation is over the atomic electrons. The symbol $\hat{\boldsymbol{\varepsilon}}(0)$ indicates that the co-ordinates are set to zero after acting on the electric field with the gradient operator $\boldsymbol{\nabla}$. This expression can be written

$$\Delta\hat{H}_2 = \frac{1}{2}\boldsymbol{\nabla} \cdot \boldsymbol{Q} \cdot \hat{\boldsymbol{\varepsilon}}(0) \tag{2.208}$$

Here $\boldsymbol{Q}$ is the quadrupole moment tensor defined by the expression

$$\boldsymbol{Q} = q\sum_j \boldsymbol{r}_j \boldsymbol{r}_j \tag{2.209}$$

If we consider a plane wave polarized in the x direction and travelling along the z co-ordinate, then

$$\Delta\hat{H}_{E2} = \frac{1}{2}q\frac{x_z}{E_0}\left(\frac{\partial E_x}{\partial z}\right) \equiv \frac{1}{2}qk_x xy \tag{2.210}$$

This last equation can also be derived by directly evaluating the terms in the expansion (2.107) using the Heisenberg equation of motion to transform the momentum.

The B_{12} coefficients for magnetic dipole and electric quadrupole radiation can now be found by substituting the expressions (2.206) and (2.208) into Equations (2.201) and (2.200) and averaging over all directions. The value for magnetic dipole radiation is

$$B_{12}^{M1} = \frac{\pi\mu_0}{3\hbar^2}|\langle\phi_2|\boldsymbol{\mu}|\phi_1\rangle|^2 \tag{2.211}$$

and for electric quadrupole radiation

$$B_{12}^{E2} = \frac{\pi q^2 \omega_{12}^2}{40\varepsilon_0 \hbar^2 c^2} |\langle \phi_2 | \boldsymbol{Q} | \phi_1 \rangle|^2 \tag{2.212}$$

Correspondingly, the A coefficients are

$$A_{21}^{M1} = \frac{g_1 \mu_0 \omega_{12}^3}{3 g_2 \pi \hbar c^3} |\langle \phi_2 | \boldsymbol{\mu} | \phi_1 \rangle|^2 \tag{2.213}$$

and

$$A_{21}^{E2} = \frac{g_1 q^2 \omega_{12}^5}{40 g_2 \pi \varepsilon_0 c^5} |\langle \phi_2 | \boldsymbol{Q} | \phi_1 \rangle|^2 \tag{2.214}$$

Unlike the electric dipole case, the operators $\boldsymbol{\mu}$ and $\boldsymbol{Q}$ connect states of the same parity, and it is possible for both types of radiation to contribute to the same transition. Rough estimates can now be made of the relative strengths of the different types of radiative decay. The ratio of the A coefficients for electric dipole and magnetic dipole radiation is

$$\frac{A^{M1}}{A^{E1}} \simeq \frac{\varepsilon_0 \mu_0}{(2m)^2} \left| \frac{\boldsymbol{L}}{\boldsymbol{R}} \right|^2 \simeq \left(\frac{\hbar}{2 m a_0 c} \right)^2 = \left(\frac{\alpha}{2} \right)^2 \tag{2.215}$$

where α is the fine structure constant, approximately 1/137. Similarly, the ratio of transition rates for electric quadrupole and electric dipole radiation is

$$\frac{A^{E2}}{A^{E1}} \simeq \frac{3}{40} \frac{\omega_{12}^2}{c^2} \left| \frac{\boldsymbol{Q}}{\boldsymbol{R}} \right|^2 \simeq \frac{3}{40} \left(\frac{\omega a_0}{c} \right)^2 = \frac{3}{40} \left(\frac{\hbar \omega}{\alpha m c^2} \right)^2 \tag{2.216}$$

which is of the order of 10^{-8} for visible transitions. Magnetic dipole and electric quadrupole transitions are therefore considerably hindered relative to electric dipole transitions. They occur when decay by $E1$ radiation is prevented or considerably hindered by selection rules. Since the lifetime associated with a strong visible transition is approximately 10^{-8} s, atomic levels which decay predominantly by $M1$ or $E2$ radiation have lifetimes of the order of milliseconds or more, and are therefore metastable states.

2.12. Selection rules and transition rates

In principle, the transition rates for electric dipole radiation can now be calculated for any atom. These rates depend on the details of the atomic wave functions via the matrix elements of the electric dipole operator. Equation (2.192) can be written

$$A_{21} = \frac{q^2\omega_{12}^3}{3g_2\pi\varepsilon_0\hbar c^3}\sum_{M_1M_2}|\langle J_2M_2|\boldsymbol{R}|J_1M_1\rangle|^2$$
$$\equiv \frac{q^2\omega_{12}^3}{3g_2\pi\varepsilon_0\hbar c^3}\sum_{M_1M_2}|I_{J_2M_2,J_1M_1}|^2 \tag{2.217}$$

where g_2 is the statistical weight (or M_2 degeneracy of J_2), equal to $2J_2+1$. Now the simplest cases to consider are the one electron systems, H, He^+, Li^{++}, etc, for which the wave functions were given in Chapter 1. To calculate the dipole matrix elements the vector $\boldsymbol{r}$ is first conveniently expressed in terms of the unit vectors

$$\boldsymbol{e}^1 = (-1/\sqrt{2})(\hat{\boldsymbol{x}}+i\hat{\boldsymbol{y}}) \qquad \boldsymbol{e}^{-1} = (1/\sqrt{2})(\hat{\boldsymbol{x}}-i\hat{\boldsymbol{y}}) \tag{2.218}$$

and

$$\boldsymbol{e}^0 = \hat{\boldsymbol{z}} \tag{2.219}$$

where $\hat{\boldsymbol{x}}$, $\hat{\boldsymbol{y}}$ and $\hat{\boldsymbol{z}}$ are unit vectors along the cartesian axis. (The significance of this choice of basis will soon become apparent.) With these definitions, $\boldsymbol{r}$ is

$$\boldsymbol{r} = (-1/\sqrt{2})r\sin\theta\exp(i\phi)\boldsymbol{e}^1 + (1/\sqrt{2})r\sin\theta\exp(-i\phi)\boldsymbol{e}^{-1} + r\cos\theta\boldsymbol{e}^0 \tag{2.220}$$

Here r, θ and ϕ are the normal spherical co-ordinates of the vector $\boldsymbol{r}$. The matrix elements, I, will then have three separate terms corresponding to each of the unit vectors, which are

$$I^{\pm 1}_{n_2l_2m_2,n_1l_1m_1} = N_0\int_0^\infty R^*_{n_2l_2}\,r\,R_{n_1l_1}\,r^2\,dr\int_0^\pi P_{l_2}^{m_2*}(\mp\sin\theta)P_{l_1}^{m_1}\sin\theta\,d\theta$$
$$\times\int_0^{2\pi}\exp(i\phi(m_2-m_1\pm 1))\,d\phi \tag{2.221}$$

and

$$I^{0}_{n_2l_2m_2,n_1l_1m_1} = N_0\int_0^\infty R^*_{n_2l_2}\,r\,R_{n_1l_1}\,r^2\,dr\int_0^\pi P_{l_2}^{m_2*}\cos\theta P_{l_1}^{m_1}\sin\theta\,d\theta$$
$$\times\int_0^{2\pi}\exp(i\phi(m_2-m_1))\,d\phi \tag{2.222}$$

The spherical wave functions, Y_l^m, have been expressed as a product of a Legendre polymomial, dependent on $\cos\theta$, and an exponential term dependent on ϕ. N_0 is the normalization constant for these functions and will depend on l_2, l_1, m_2 and m_1. These two equations, (2.221) and (2.222), contain the information concerning the selection rules for electric dipole radiation, for single electron transitions. Notice, however, that the electron spin wave functions are not included because the interaction contains no magnetic terms. This, in fact, conceals the selection rule that the spin magnetic quantum number is conserved during the transition. Spin flip transitions are only possible for magnetic multipole interactions.

Consider, firstly, the selection rules for m (the magnetic orbital quantum number), which are contained in the ϕ integrals. $I^{\pm 1}$ vanish unless Δm, which equals $m_2 - m_1$, is ± 1, and I^0 vanishes unless Δm equals zero. If the z axis is defined by a weak magnetic field, then the radiation emitted (or absorbed†) along this axis will have its polarization described by the unit vectors $\boldsymbol{e}^1$ and $\boldsymbol{e}^{-1}$. Note carefully that $\boldsymbol{r}$ is the position vector with reference to the atom and it does not refer directly to the radiation field. However, the interaction is more appropriately defined by the matrix elements of $\hat{\boldsymbol{\varepsilon}}.\boldsymbol{r}$, where $\hat{\boldsymbol{\varepsilon}}$ is the unit vector of the radiation field, so we can choose $\hat{\boldsymbol{\varepsilon}}$ to have the same basis as $\boldsymbol{r}$. By writing down the complete time dependence of the radiation of these as given by Equation (2.93), it can be seen that $\boldsymbol{e}^1$ defines a left hand circularly polarized wave and $\boldsymbol{e}^{-1}$ a right hand circularly polarized wave. Radiation emitted normal to the magnetic axis will have plane polarized components, from I^0 and $I^{\pm 1}$. This is considered in more detail in the next section.

The selection rules for orbital angular momentum, l, are contained in the central integral of Equations (2.221) and (2.222). These are only non-zero, in the first instance, when Δm is given by the previous analysis, that is $\Delta m = \pm 1, 0$. Most importantly, the integrals also vanish unless $l_2 = l_1 \pm 1$. From these simple analyses it is possible to draw conclusions concerning the properties of photons. Radiation emitted along the z axis cannot have a component of 'orbital' angular momentum in this direction, implying that the change in momentum is brought about by the intrinsic angular momentum of the photon. We thus assign a spin equal to $1\hbar$ to the photon, and the component of spin in the direction of propagation defines the helicity. A photon with helicity $+1$, is left circularly polarized whilst a helicity of -1, is right circularly polarized. Because of the transverse nature of the electromagnetic wave, helicity zero is excluded; plane polarized waves consist of equal numbers of the two photon states. If the spin of the electron is now introduced, a further selection rule results from the conservation of angular momentum. This is that, $\Delta j = j_2 - j_1$, must be either 0 or ± 1, except for transitions from $j_1 = 0$ to $j_2 = 0$, which are clearly not possible.

Next, consider the parity selection rules. Since the parity operator, P, anticommutes with the electric dipole moment it follows that the matrix elements (2.221) and (2.222) vanish unless the two states have opposite parity. The electric

† Absorption and emission will differ in the sign of Δm for the same circular polarization.

Table 2.1. Electric dipole transition rates in atomic hydrogen

Initial state	Final state 1*s*		2*s*		2*p*	
	$A(s^{-1})$	λ(nm)	$A(s^{-1})$	λ(nm)	$A(s^{-1})$	λ(nm)
2*p*	$6{\cdot}28 \times 10^8$	122	—	—	—	—
3*s*	—	—	—	—	$6{\cdot}31 \times 10^6$	656
3*p*	$1{\cdot}67 \times 10^8$	103	$2{\cdot}24 \times 10^7$	656	—	—
3*d*	—	—	—	—	$6{\cdot}46 \times 10^7$	656
4*s*	—	—	—	—	$2{\cdot}58 \times 10^6$	486
4*p*	$6{\cdot}82 \times 10^7$	97	$9{\cdot}67 \times 10^6$	486	—	—
4*d*	—	—	—	—	$2{\cdot}06 \times 10^7$	486

quadrupole and magnetic dipole operarors commute with P, so these transitions will involve no parity change.

If the selection rules outlined previously are obeyed, the two angular integrals can be given explicitly in terms of the quantum numbers, l_2, l_1, m_2 and m_1. These values, as well as more general ones involving the coupling of angular momenta for many electron systems, can be found in detailed texts concerned with the theory of atomic spectra. Since these angular 'coupling' factors are of order unity, the transition rates are principally determined by the radial integrals. For hydrogen-like systems they can be calculated exactly, and as an example Table 2.1 gives the rate coefficients for electric dipole transitions in atomic hydrogen. Where experimental data exists they are in good agreement. As a general rule, the square of the radial integrals is largest when the two states have the same principle quantum number, and they fall off rapidly as the difference, $\Delta n = n_2 - n_1$, is increased. This means that the B coefficients, describing the absorption of radiation from an atomic ground state to a high n state ($n > 20$), are very small. Even so, Rydberg atoms can be produced using selective laser excitations, a possibility which is brought about by the extremely high spectral intensity of laser sources.

Transition rates can also be calculated for the alkali metals in a similar fashion. Here, the angular integrals will be identical, but the radial integrals will involve wave functions of the central field approximation. The simplest approximation is to replace the atomic number Z, in the radial wave function, with an effective charge $Z_{\text{eff}}(n, l)$, which depends on the single particle state. Values of Z_{eff} can be obtained from either a Hartree–Fock or Thomas–Fermi analysis. When n is large, the orbits are non-penetrating, and the hydrogenic wave functions can be used (i.e. $Z_{\text{eff}} = 1$).

When more than one valence electron is involved the calculation of the matrix elements becomes quite complicated. If the wave functions are best described within the framework of LS coupling, the transition rates for electric dipole radiation are proportional to

$$\sum_{M_1 M_2} |\langle \alpha_2 L_2 S_2 J_2 M_2 | \boldsymbol{R} | \alpha_1 L_1 S_1 J_1 M_1 \rangle|^2 \tag{2.223}$$

where α is used to account for the configuration and any extra variables needed to distinguish the states. Selection rules can be evaluated from these elements, using general equations for the coupling of angular momentum. The selection rules are

$$\Delta L = 0, \pm 1; \qquad \text{not } L_2 = 0 \text{ to } L_1 = 0 \tag{2.224}$$

$$\Delta S = 0 \tag{2.225}$$

$$\Delta M_L = 0, \pm 1 \tag{2.226}$$

$$\Delta J = 0, \pm 1; \qquad \text{not } J_2 = 0 \text{ to } J_1 = 0 \tag{2.227}$$

$$\Delta M = 0, \pm 1 \tag{2.228}$$

and

$$P_2 = -P_1 \tag{2.229}$$

where P_1 and P_2 are the parities of the two states. The most probable type of transition is when one electron changes its orbital angular momentum by one unit, in keeping with the single electron selection rules. If there is strong configuration mixing then two electron jumps can take place for a single photon emission (or absorption) but these are relatively uncommon. Transition rates between any two levels are therefore weaker than the corresponding single particle transition in the one electron system. In general this is because the total electric dipole strength is distributed over many more levels. For example, the two electron configuration (*nsnd*) might decay to the (*nsnp*) configuration by a single electron jump from $l_2 = 2$ to $l_2 = 1$. The transitions are therefore between the levels of the two triplet terms ^3D and ^3P, which are allowed by the J selection rule. These are $^3D_3$–$^3P_2$, $^3D_2$–$^3P_3$, $^3D_2$–$^3P_2$, $^3D_1$–$^3P_2$, $^3D_1$–$^3P_1$ and $^3D_1$–$^3P_0$. Had we chosen a configuration with many terms, the number of allowed transitions would have been much greater. The relative intensity of transitions between one fine structure multiplet and another depend on the coupling of the angular momenta involved. Details can be found in more specialized texts.

Similar selection rules apply for electronic transitions in molecules when the spin-orbit coupling is not too strong. Equations (2.227) to (2.229) still apply, whatever the circumstances, but other selection rules depend on the coupling between the rotational motion and the intrinsic electronic angular momentum. Selection rules, applying for Hund's cases (a) and (b), which cover many examples, are that $\Delta S = 0$ and $\Delta\Lambda = 0$ or ± 1. The first rule means that only states of the same multiplicity combine, whilst the second rule is analogous to the L selection (2.224). Notice that Equations (2.227) to (2.229) are *rigid* selection rules for dipole radiation and do not depend on the coupling schemes involved.

2.13. Angular distributions of the radiation

The transition probability, P, for emitting a photon of polarization, $\hat{\boldsymbol{\varepsilon}}$, into the small solid angle $d\Omega$, is given by the expression

$$P(\hat{\boldsymbol{\varepsilon}})\, d\Omega = \frac{q^2 \omega_{12}^3}{g_2\, 4\pi^2 \varepsilon_0\, \hbar c^3} \sum_{M_1 M_2} |\langle J_2 M_2 | \hat{\boldsymbol{\varepsilon}}.\boldsymbol{R} | J_1 M_1 \rangle|^2\, d\Omega \tag{2.230}$$

This can be used to calculate the total transition probability (2.217) by integrating over a 4π solid angle and noting that the average value of $(\hat{\boldsymbol{\varepsilon}}\,.\,\boldsymbol{R})$ is $\boldsymbol{R}/3$. Sometimes this equation is written with a factor 2 in the denominator if the two polarizations for each mode (two directions) are separated out. $\hat{\boldsymbol{\varepsilon}}$ can be written, as previously for $\boldsymbol{r}$, in terms of its spherical components, namely

$$\varepsilon^{+1} = \frac{-1}{\sqrt{2}}(\varepsilon_x + i\varepsilon_y) \tag{2.231}$$

$$\varepsilon^{-1} = \frac{1}{\sqrt{2}}(\varepsilon_x - i\varepsilon_y) \tag{2.232}$$

and

$$\varepsilon^0 = \varepsilon_z \tag{2.233}$$

where ε_x, ε_y and ε_z are the cartesian components of $\hat{\boldsymbol{\varepsilon}}$. The transition probability can then be separated out into three components, in the same manner as Section 2.12. For simplicity of notation, assume that the transitions involve single electrons, then the three components of P, which correspond to $\Delta m = +1$, $\Delta m = -1$, and $\Delta m = 0$, are

$$P^{\pm 1} = C_0 |\varepsilon^{\pm 1 *}|^2 \sum_{\Delta m = \pm 1} |I^{\pm 1}|^2 \tag{2.234}$$

$$P^0 = C_0 |\varepsilon^{0 *}|^2 \sum_{\Delta m = 0} |I^0|^2 \tag{2.235}$$

where

$$C_0 = \frac{q^2 \omega_{12}^3}{g_2\, 4\pi^2 \varepsilon_0\, \hbar c^3} \tag{2.236}$$

and I is given by the expressions (2.221) and (2.222). The angular part of the

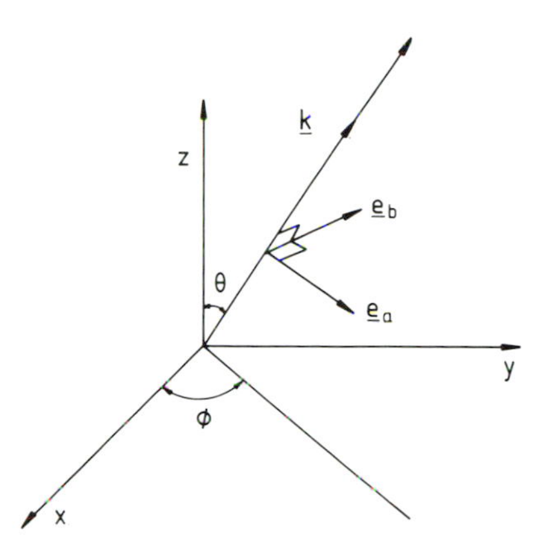

Figure 2.12. Co-ordinate systems and vectors $\hat{e}_a$ and $\hat{e}_b$ used to describe the polarization of an electromagnetic wave travelling in the $\boldsymbol{k}$ direction.

distribution is contained in the spherical components of $\hat{\boldsymbol{\varepsilon}}$ which vary as the observation angle is changed. To obtain the angular distribution the expressions (2.234) and (2.235) are written in terms of two unit vectors $\hat{\boldsymbol{e}}_a$ and $\hat{\boldsymbol{e}}_b$ which are normal to the propagation (observation) direction $\hat{\boldsymbol{k}}$. $\hat{\boldsymbol{e}}_a$, $\hat{\boldsymbol{e}}_b$ and $\hat{\boldsymbol{k}}$ form a right handed orthogonal set of vectors as shown in Figure 2.12, with $\hat{\boldsymbol{e}}_b$ chosen to lie in the xy plane. The cartesian components of these two vectors are

$$e_{az} = -\sin\theta \quad e_{ax} = \cos\theta\cos\phi \quad e_{ay} = \cos\theta\sin\phi \tag{2.237}$$

and

$$e_{bz} = 0 \quad e_{bx} = -\sin\phi \quad e_{by} = \cos\phi \tag{2.238}$$

The angular distribution for $\Delta m = 0$ (or $\Delta M = 0$) transitions, with polarization defined only by $\hat{\boldsymbol{e}}_a$ (transverse linear polarization) is then obtained by substituting the first expression of (2.237) into (2.235). Thus

$$P^0_{e_a} = C_0 \sin^2\theta \sum_{\Delta m=0} |I^0|^2 \tag{2.239}$$

Similarly the contribution to $\Delta m = -1$ with polarization defined by $\hat{\boldsymbol{e}}_a$ is

$$P^{-1}_{e_a} = \frac{C_0}{2}|\cos\phi + i\sin\phi|^2\cos^2\theta \sum_{\Delta m=-1} I^{-1} \tag{2.240}$$

and the contribution to $\Delta m = -1$ with polarization defined by $\hat{\boldsymbol{e}}_b$ is

$$P^{-1}_{e_b} = \frac{C_0}{2}|-\sin\phi + i\cos\phi|^2 \sum_{\Delta m=-1} I^{-1} \tag{2.241}$$

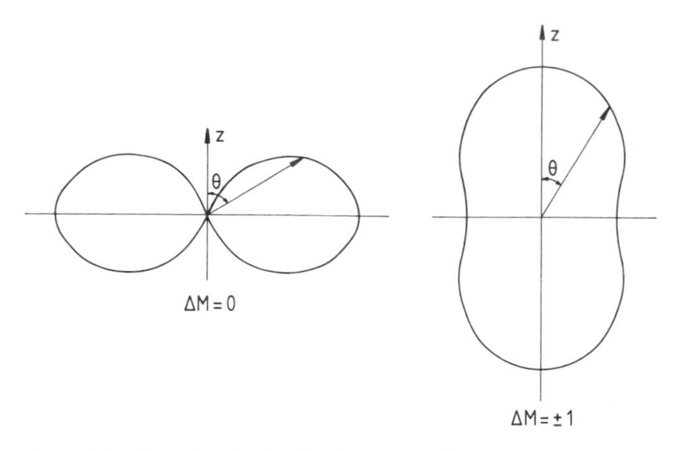

Figure 2.13. Polar plots of angular distributions for $\Delta M = 0$ and $\Delta M = \pm 1$ Zeeman components.

The total contribution from both states of polarization is therefore

$$P^{-1}_{e_a+e_b} = \frac{C_0}{2}(1+\cos^2\theta) \sum_{\Delta m = -1} I^{-1} \tag{2.242}$$

When θ is zero the vectors $\hat{e}_a$ and $\hat{e}_b$ describe a right hand circularly polarized wave propagating along the z direction, whilst for $\theta = \pi/2$, $P^{-1}_{e_a}$ vanishes and the wave is linearly polarized in the xy plane. Similarly, the total probability for $\Delta m = +1$ transition from both states of polarization is

$$P^{+1}_{e_a+e_b} = \frac{C_0}{2}(1+\cos^2\theta) \sum_{\Delta m = +1} I^{+1} \tag{2.243}$$

and the polarization will alter from left circular to plane as the observation direction is changed from $\theta = 0$ to $\theta = \pi/2$. The angular distributions are shown in Figure 2.13.

In the absence of a magnetic field all the M substates are degenerate and if these are equally populated then the angular distribution, obtained from summing the three terms, is isotropic and unpolarized. However, if the excited level is populated by the absorption of a polarized light source, then the magnetic substates will not be equally populated and the angular distribution will be, in general, anisotropic. The exact form of this anisotropy can be written in terms of angular momentum coupling coefficients, and can be found in other tests. In a magnetic field the M degeneracy is removed, whereupon the transitions corresponding to $\Delta m = \pm 1$ (σ polarization) can be distinguished from those corresponding to $\Delta m = 0$ (π polarization). The magnetic field vector then defines the z axis with the angular distributions and polarizations given by the previous expressions.

Bibliography

Bransden, B.H. and Goachain, C.J., 1983, *Physics of Atoms and Molecules* (London: Longman).
Condon, E.U. and Shortley, G.H., 1963, *The Theory of Atomic Spectra* (Cambridge University Press).
Corney, A., 1977, *Atomic and Laser Spectroscopy* (Oxford University Press).
Demtroder, W., 1981, *Laser Spectroscopy* (Berlin: Springer-Verlag).
Heitler, W., 1960, *The Quantum Theory of Radiation* (3rd edition) (Oxford University Press).
Landau, L.D. and Lifshitz, E.M., 1960, *Electrodynamics of Continuous Media* (Oxford: Pergamon Press).
Loudon, R., 1981, *The Quantum Theory of Light* (2nd edition) (Oxford University Press).
Slater, J.C. and Frank, N.H., 1947, *Electromagnetism* (New York: McGraw-Hill).
Thorne, A.P., 1974, *Spectrophysics* (London: Chapman and Hall).
Woodgate, G.K., 1980, *Elementary Atomic Structure* (2nd edition) (Oxford University Press).

CHAPTER 3
light amplification

3.0. Introduction

The previous chapter considered the interaction of an electromagnetic wave with two completely isolated energy levels. Under steady state conditions, and for non-degenerate levels, the number of atoms in the excited state, N_2, cannot exceed the number in the ground state, N_1, and a propagating wave is always attenuated. Now we consider a more realistic situation where, first, the two levels are connected to other levels by electromagnetic transitions, and, second, the lower level need not be the ground state. Furthermore, energy levels can be populated or depopulated by other mechanisms (such as electron or atomic collisions) which need to be considered. In any circumstance, if the number of atoms N_2, in the upper level E_2, exceed the number N_1, in the lower level E_1, then a wave propagating with an angular frequency $(E_2 - E_1)/\hbar$, can be amplified. This condition, in which N_2 (or $g_1 N_2$ when the degeneracies of the levels are considered), is greater than N_1 (or $g_2 N_1$) is known as a population inversion because it is completely opposite to that found in an isolated system in thermal equilibrium. It is the fundamental prerequisite for any laser whatever the type.

In this chapter the basics of laser theory will be studied, starting with the equations for the amplification of a travelling wave, and finishing with a simplified theory for a standing wave oscillator. This section also serves as a brief introduction to saturated absorption spectroscopy. The way in which the population inversion is produced depends on the type of laser. These are described in detail in subsequent chapters but a brief description of a standing wave laser is given in order to provide the necessary background information.

3.1. General equation for the absorption coefficient in terms of the level populations

Consider a steady electromagnetic wave propagating through a medium containing excited atoms and molecules. If the frequency of oscillation corresponds to the transition energy between two levels E_1 and E_2 then the wave amplitude will be altered as it traverses the medium. When $g_2 N_1$, where N_1 refers to the lower level, is greater than $g_1 N_2$ the wave is absorped whilst for $g_2 N_1$ less than $g_1 N_2$, the wave is amplified. To show this, we calculate the absorption coefficient in terms of the Einstein coefficients and the level populations. Firstly, the intensity variation in the direction of propagating, z, is related to the energy density by means of the equation

$$c\,W(z) = \eta\,I(z) \tag{3.1}$$

where η is the refractive index, and both W and I refer to averages taken over one cycle of the field oscillation. (This equation is obtained by considering the balance of energy in a thin slice of the medium.) If we ignore for the moment any external influences and assume, as in Chapter 2, that the two levels are completely isolated, then we can write down the steady state condition for the numbers of atoms in the two levels by equating the rate at which atoms decay by spontaneous and stimulated emission to the rate at which they are produced by absorption. This equation is

$$N_2(z)A = ((g_2/g_1)N_1(z) - N_2(z))B\,W(\omega, z)/\eta^2 \tag{3.2}$$

where for simplicity of notation we have put $B_{21} = B$. Notice that this includes the factor η^2, since the definitions of A and B in Chapter 2 were calculated in a vacuum ($\varepsilon = 1$). Now, this equation does not take into account the variation of absorption and stimulated emission as a function of the incident light frequency. If the absorption coefficient for a monochromatic light beam is required then the effects of line broadening need to be considered. This can be achieved for homogeneous broadening by multiplying the Einstein B coefficient by a term $f(\omega)\,d\omega$, where $f(\omega)$ represents a suitably normalized frequency response which will have, in general, a Lorenzian profile. To obtain the total number of atoms N_2, Equation (3.2) could be integrated over the total line shape, provided that the frequency dependence of the energy density is known. In the case of black-body radiation the change in W is negligible over the line profile so the integral will reduce to (3.2) (or Equation (2.124) with (dN_2/dt) equal to zero) provided that $f(\omega)$ is correctly normalized. In most cases we are interested in the response from a monochromatic source where the intensity is represented by a delta function, and equation (3.2) becomes

$$N_2(z)A = ((g_2/g_1)N_1(z) - N_2(z))B\,W_\omega(z)f(\omega)/\eta^2 \tag{3.3}$$

where, now, W_ω is the total energy per unit volume. Using the conservation equation, $N_0 = N_1 + N_2$, N_2 can be written in terms of the total number N_0 and the Einstein coefficients. This equation is identical in form to (2.131) except that the B coefficients are now scaled by the factor $f(\omega)$. It is directly equivalent to that

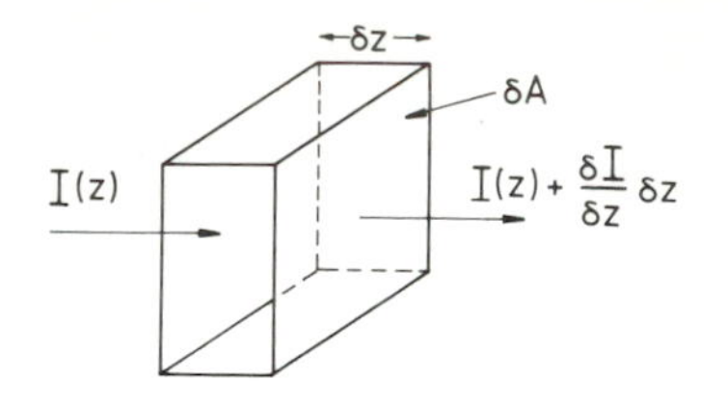

Figure 3.1. The change in intensity of a beam as it passes through a thin slice of the medium. $I(\omega)\,d\omega$ is the intensity for frequencies between ω and $\omega + d\omega$.

found using the coupled oscillator equations† of Chapter 2 (Equations (2.168) and (2.124)) if the frequency dependent factor has a normalized Lorenzian profile

$$f(\omega) = \frac{(\gamma_t/\pi)}{(\omega_{12} - \omega)^2 + \gamma_t^2} \tag{3.4}$$

with γ equal to the half width at half maximum height.

Next, consider a thin slice of the medium at a distance z along the path of the propagating beam, as shown in Figure 3.1. If this has a small area δA, the rate at which the energy changes in the beam, for a range of frequencies from ω to $\omega + d\omega$, is

$$(I(\omega, z) - I(\omega, z + dz))\,\delta A\,d\omega = (-\,\partial I(\omega, z)/\partial z)\,dz\,\delta A\,d\omega \tag{3.5}$$

where $I(\omega, z)$ is the *spectral* intensity of the propagating beam. This change can be equated to the difference between the rate at which energy is absorbed from the beam and the rate at which it is put back into the beam by stimulated emission.

For a small range of frequencies $d\omega$, this is just

$$((g_2/g_1)n_1(z) - n_2(z))B\,W(z, \omega)\hbar\omega\,f(\omega)\,d\omega\,\delta A\,dz/\eta^2 \tag{3.6}$$

Here n_1 and n_2 are the number densities of atoms and $f(\omega)$ is normalized according to the integral

$$\int_0^\infty f(\omega)\,d\omega = 1 \tag{3.7}$$

which ensures that the Einstein coefficients are in accord with the previous analysis. Combining (3.5), (3.6) and (3.1) gives

$$\frac{\partial I(\omega, z)}{\partial z} = \left[n_2(z) - \left(\frac{g_2}{g_1}\right)n_1(z)\right]\frac{B\hbar\omega\,f(\omega)\,I(\omega, z)}{c\eta} \tag{3.8}$$

† For this it is required that $BW/A = \pi\Omega^2/4\gamma_t$. This can be established using the definitions of the previous chapter.

It is important to understand that this equation applies even when the two levels are not isolated, since it does not consider any balance between n_1 and n_2. For the two level system, under no external influences, the equilibrium balance is given by (3.3), and the changes of n_1 and n_2 are predicted by the Equations (2.124) to (2.131). In this case the alteration of intensity across the slab must be related to the rate at which energy is removed by spontaneous emission. Equation (3.8) will also be true if we are considering the total intensity†, I_ω, in a monochromatic beam. I_ω will then be an integral of a spectral intensity represented by a delta function.

The solution to (3.8) when n_1 and n_2 are independent of z (or I) is simply

$$I = I_0 \exp(-Kz) \tag{3.9}$$

where I_0 is the intensity at $z = 0$, and the absorption coefficient K is defined by the equation

$$K(\omega) = -(n_2 - (g_2/g_1)n_1)B\hbar\omega\, f(\omega)/c\eta \tag{3.10}$$

The more usual form of this equation is obtained by expressing the B_{21} coefficient in terms of A_{21} coefficient using the formula (2.24), but now modified to include the refractive index. Thus,

$$K(\omega) = -(n_2 - (g_2/g_1)n_1)\frac{\pi^2 c^2 A_{21}}{4\omega^2\eta^2} f(\omega) \tag{3.11}$$

which can be written in terms of the wavelength, λ, and the frequency, ν, as

$$K(\lambda) = -(n_2 - (g_2/g_1)n_1)\lambda^2 A_{21}\,\alpha(\nu)/8\pi\eta^2 \tag{3.12}$$

where $\alpha(\nu)$ is the homogeneously broadened line shape in frequency space. α is related to f by the simple relation, $\alpha = (d\omega/d\nu)f$. Now, it is fairly easy to see that for negative values of K, when n_2 is greater than $(g_2/g_1)n_1$, the intensity, as defined by Equation (3.8), grows exponentially, whilst for n_2 less than $(g_2/g_1)n_1$, the wave decreases in amplitude. Equation (3.12) for the absorption coefficient, which is used to calculate the gain of a laser, is one of the most important equations in laser physics. In Sections 3.2 and 3.3 values for this gain are calculated in terms of the pumping rates for homogeneous and inhomogeneous transitions.

Before studying the laser case in detail, it is worthwhile analysing the situation for a two level system with no external interactions other than the propagating light beam. Here the equilibrium values for n_2 and n_1 are given by expression (3.3), and the conservation equation $n_2 + n_1 = n_0$. Combining these

† $I(\omega)$ represents a spectral intensity whereas I_ω is a total intensity in a monochromatic beam.

two gives

$$((g_2/g_1)n_1 - n_2) = \frac{n_0(g_2/g_1)A_{21}}{(A + (1 + (g_2/g_1))BW f/c\eta^2}$$

$$= \frac{n_0(g_2/g_1)A_{21}}{(A + (1 + (g_s/g_1))BI f/c\eta^2} \qquad (3.13)$$

which can then be substituted into (3.3) to yield

$$\frac{\partial I}{\partial z}\left[\frac{g_1}{g_2} + \left(1 + \frac{g_1}{g_2}\right)\left(\frac{BIf}{A_{21}\,c\eta}\right)\right] = -n_0\, B\hbar\omega f(\omega) I/c\eta \qquad (3.14)$$

When the beam intensity is weak, so that fBI is much smaller than $A_{21}\,c\eta$, the second term in the square brackets can be ignored and the intensity decreases exponentially with an absorption coefficient

$$K(\omega) = n_0(g_2/g_1)B\hbar\omega\, f(\omega)/c\eta \qquad (3.15)$$

For a very intense beam the second term is larger, and expression (3.14) can be approximated to

$$\frac{\partial I}{\partial z} = \frac{-n_0\, A_{21}\, \hbar\omega}{(1 + (g_1/g_2))} \qquad (3.16)$$

Integrating this gives the linear expression

$$I_\omega(z) = I_0 - \left(\frac{n_0\, A_{21}\, \hbar\omega}{(1 + (g_1/g_2))}\right) z \qquad (3.17)$$

for the variation in intensity. Notice that in both cases the beam intensity decreases as a function of z. In the second case the absorption is only determined by the A coefficient, whereas, for a weak beam, it is determined solely by the B coefficient. This is quite realistic physically since for the second case (Equation 3.16) the transition is saturated, and the energy can only be extracted at a rate determined by the decay of the excited atoms.

3.2. Population inversion: general equation for amplification for homogeneous transitions

Next consider the case of a monochromatic wave propagating in a medium where the levels are not isolated but can be populated by alternative means. A

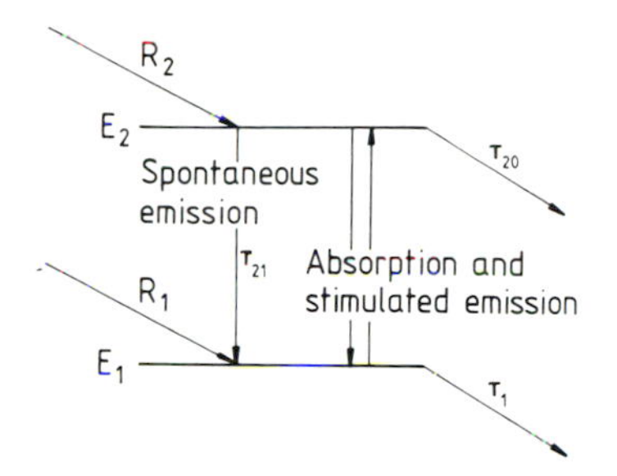

Figure 3.2. Diagram illustrating the various constants involved in calculating the level populations. Both E_2 and E_1 are above the ground state, which effectively provides a reservoir of atoms that can be pumped into the excited states.

general scheme for this is shown in Figure 3.2. We suppose that the level E_2 is populated at a rate R_2 and can decay to the lower level with a lifetime $\tau_{21} = 1/A_{21}$. This level can also be depopulated by electromagnetic decay to other levels (or by relation processes) with a partial lifetime of τ_{20}. R_2 is meant to include all the processes which can populate E_2, for example, electron impact, collision excitation or decay from higher levels. The specific mechanism(s) will depend on the type of laser, and these are discussed in detail later in the book. Rate equations can now be given relating n_2 and n_1 to the various decay paths. For homogeneous broadening these are

$$\frac{dn_2}{dt} = R_2 - \frac{n_2}{\tau_{20}} - \frac{n_2}{\tau_{21}} - \frac{BW_\omega f(\omega)(n_2 - (g_2/g_1)n_1)}{\eta^2} \tag{3.18}$$

and

$$\frac{dn_1}{dt} = R_1 - \frac{n_1}{\tau_1} + \frac{n_2}{\tau_{21}} + \frac{BW_\omega f(\omega)(n_2 - (g_2/g_1)n_1)}{\eta^2} \tag{3.19}$$

Here R_2 and R_1 are the rates for the change in the number density of atoms. Now, previously, we considered the case where R_2, R_1, τ_{20} and τ_{10} were all zero, and for the steady state solution, this gave Equation (3.3) ($dn_1/dt = dn_2/dt = 0$). Again considering the steady state solutions, which must apply for a continuous wave (CW) laser, we obtain two equations for R_2 and R_1 in terms of the intensity via Equation (3.1) and the populations n_2 and n_1. These steady state solutions are

$$R_2 = n_2\left[\frac{1}{\tau_{20}} + \frac{1}{\tau_{21}} + \frac{BIf}{\eta c}\right] + n_1\left[-\frac{BIf(g_2/g_1)}{\eta c}\right] \tag{3.20}$$

and

$$R_1 = n_2\left[-\frac{1}{\tau_{21}} - \frac{BIf}{\eta c}\right] + n_1\left[\frac{1}{\tau_1} + \frac{BIf(g_2/g_1)}{\eta c}\right] \tag{3.21}$$

These two equations are now used to obtain the population differences needed to calculate the absorption coefficient in Equation (3.10). In doing this we are assuming that R_2 and R_1 are essentially fixed by the pumping power and are not altered by the lasing action. This is only true if the ground state is effectively a large reservoir of atoms so that n_0 is always larger than n_1 or n_2. When the pumping is very hard, or E_1 is the ground state, then this will not apply, and conservation of population must be considered. This will be treated in detail for 3 and 4 level systems later on in the book. Now, in laser physics, it is more usual to discuss the amplification of the beam in terms of a gain coefficient $\alpha(\omega)$, where $\alpha(\omega) = -K(\omega)$. For positive values of α the beam intensity increases in amplitude and the small signal gain is given by the expression

$$G = \exp(\alpha z) \tag{3.22}$$

where the intensity is

$$I = I_0 G \tag{3.23}$$

and z is the distance of propagation. Notice again that (3.20) and (3.21) are for a monochromatic wave of intensity I_ω and angular frequency ω. The first part of Equation (3.8) can be written

$$n_2 - (g_2/g_1)n_1 = \frac{R_2\tau_2(1-(g_2/g_1)(\tau_1/\tau_{21})) - R_1(g_2/g_1)\tau_1}{1+(\sigma(\omega)I_\omega/\hbar\omega)((g_2/g_1)\tau_1 + \tau_2 - (g_2/g_1)(\tau_1\tau_2/\tau_{21}))} \tag{3.24}$$

where the total lifetime of the upper state, τ_2, is given by

$$\frac{1}{\tau_2} = \frac{1}{\tau_{20}} + \frac{1}{\tau_{21}} \tag{3.25}$$

The B coefficient has now been incorporated into a single atom cross section, $\sigma(\omega)$, by means of the relation

$$\sigma(\omega) = Bf(\omega)\hbar\omega/\eta c \tag{3.26}$$

so that $\sigma(\omega)$ is the cross section (in units of L^2) for stimulated emission of radiation (or absorption if B_{12} is used) at frequency ω. Equations (3.24) and (3.8) can now be used to find the coefficient $\alpha(\omega)$ at equilibrium. This is

$$\alpha(\omega) = \frac{(R_2\tau_2(1-(g_2/g_1)(\tau_1/\tau_{21})) - R_1(g_2/g_1)\tau_1)\sigma(\omega)}{1+(\sigma(\omega)I_\omega/\hbar\omega)((g_2/g_1)\tau_1 + \tau_2 - (g_2/g_1)(\tau_1\tau_2/\tau_{21}))} \tag{3.27}$$

and the intensity variation as a function of distance of propagation, z, is a solution of the differential equation

$$\frac{\partial I}{\partial z} = \alpha(I) . I \tag{3.28}$$

The gain is therefore dependent on the intensity of the beam. At low intensities, the second term on the bottom line of (3.27) can be ignored and the gain coefficient is given by the formula

$$\alpha_0(\omega) = [R_2 \tau_2(1-(g_2/g_1)(\tau_1/\tau_{21})) - R_1(g_2/g_1)\tau_1]\sigma(\omega) \tag{3.29}$$

$$= (n_2^0 - (g_2/g_1)n_1^0)\sigma(\omega) \tag{3.30}$$

Here, α_0, which is known as the small signal gain coefficient, has been written in terms of the initial populations n_2^0 and n_1^0. These are the densities that are present before the laser power is large enough to transfer significant numbers from E_2 to E_1 by stimulated emission. The small signal gain of the system in these circumstances is

$$G = \frac{I}{I_0} = \exp(\alpha_0 z) \tag{3.31}$$

At the other extreme, the second term in the denominator of (3.27) is dominant, and the variation in intensity is given by the linear equation

$$I(z) = I_0 + \left(\frac{\alpha_0 I_s}{\bar{f}}\right) z \tag{3.32}$$

where $I_s(\omega_{12})$ is the saturation parameter and $\bar{f}(\omega)$ is the Lorenzian frequency response normalized to unity at the line centre, ω_{12}. I_s is given by the formula

$$I_s = \frac{h\omega_{12}}{\sigma(\omega_{12})((g_2/g_1)\tau_1 + \tau_2 - (g_2/g_1)(\tau_1 \tau_2/\tau_{21}))} \tag{3.33}$$

and the Lorenzian, $\bar{f}$, is

$$\bar{f}(\omega) = \frac{\gamma_t^2}{(\omega_{12} - \omega)^2 + \gamma_t^2} = \frac{f(\omega)}{f(\omega_{12})} \tag{3.34}$$

The gain is now only determined by the pumping rates and lifetimes. This is analogous to the previous discussion in Section 3.1. At this point we are just adding a number of photons per unit length to the beam at an effective pumping rate R_{eff}, which is

$$R_{\text{eff}} = \frac{R_2\tau_2(1-(g_2/g_1)(\tau_1/\tau_{21})) - R_1(g_2/g_1)\tau_1}{(g_2/g_1)\tau_1 + \tau_2 - (g_2/g_1)(\tau_1\tau_2/\tau_{21})} \tag{3.35}$$

$$= \frac{\alpha_m}{\hbar\omega} \tag{3.36}$$

where α_m is the gain coefficient at saturation. Notice that the general equation for the gain coefficient (3.27) can be written in the simpler form, using (3.33), (3.29) and (3.34) as

$$\alpha(\omega) = \frac{\alpha_0(\omega)}{1+(I_\omega/I_s)\bar{f}(\omega)} \tag{3.37}$$

Figure 3.3 illustrates this variation of the gain coefficient (3.37) as a function of frequency and intensity. The absorption for all values of intensity can be obtained by substituting (3.37) into (3.28). This produces the general transcendental equation

$$\ln\left(\frac{I(z)}{I_0}\right) + \left(\frac{\bar{f}(\omega)}{I_s}\right)(I(z)-I_0) = \alpha_0 z \tag{3.38}$$

which can only be solved numerically, except for the extremes considered previously.

The gain coefficient (3.37) will be used later in the chapter to discuss the

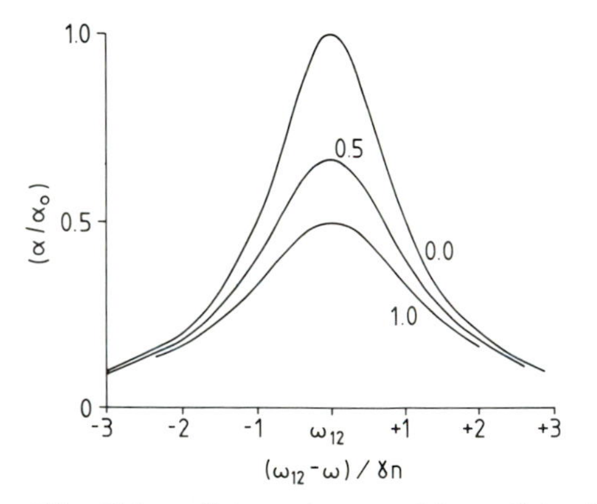

Figure 3.3. Gain coefficient, α, in terms of the small signal gain, α_0, for a homogeneously broadened transition. Values are plotted for $I/I_s = 0$, 0·5 and 1·0 where I_s is the saturation parameter. As the intensity increases the whole gain curve reduces.

amplification in a practical laser cavity. However, it is worth considering what happens to the different cavity modes for a homogeneously broadened transition. The mode whose frequency lies closest to the peak of the small signal gain curve will initially remove the most energy from the system. The gain curve will fall (see Figure 3.3) until a point is reached where the losses in the cavity can be balanced with a particular value of α and, of course, intensity, I_ω. Since the whole gain curve is affected other cavity modes will have insufficient gain to balance the losses and these will be extinguished. Thus we expect lasers with homogeneously broadened transitions to oscillate preferentially on a single cavity mode. This is a simple reflection of the homogeneous nature of the transition. For example, atoms which absorb light of a particular frequency can be transferred to E_2 and then decay back to E_1 by spontaneous or stimulated emission. These atoms are then free to absorb light of another frequency with a probability determined by the Lorenzian distribution. In other words the effect of a monochromatic source is to alter the whole of the gain curve α. This behaviour is in contrast to inhomogeneous broadening, which is discussed in the next section.

Some general observations can be made about the gain function (3.27) in relation to the fundamental lifetimes and decay rates shown in Figure 3.2. So as not to complicate the issue we will assume that the degeneracies g_1 and g_2 are both unity. For lasing action, τ_1 must be smaller than τ_{21} (since $\alpha(\omega)$ must be positive) and it is advantageous if R_2 is large whilst R_1 is small. Additionally, the most favourable conditions are also when the upper level lifetime is determined by radiative decay to the lower level, and when τ_2 is very much larger than τ_1. For many lasers this condition is true and the saturated intensity is approximate to

$$I_s = \frac{\hbar\omega_{12}}{\tau_2\,\sigma(\omega)} \tag{3.39}$$

Also, if $\tau_{21} = \tau_2 \gg \tau_1$, then Equation (3.29) becomes

$$\alpha_0(\omega) = R_2\,\tau_{21}\,\sigma(\omega) = R_2\,\tau_2\,\sigma(\omega) \tag{3.40}$$

Now we have a large pumping rate populating only the upper level which has a relatively slow spontaneous decay to the lower level. The lifetime of the lower level is short so this is depleted as fast as possible. In this condition the population of E_1 is zero to begin with ($n_1^0 = 0$) and only builds up as saturation begins. This saturation can be thought of as reducing the effective lifetime of an atom in the upper level so that ultimately it begins to compete with the lifetime τ_1. Of course, τ_{21}, which is just $1/A_{21}$, is fixed.

3.3. Gain curves for inhomogeneous transitions

The previous discussion applied only for homogeneous transitions and needs considerable modification to account for, in particular, the effects of

Doppler shifts (other causes of inhomogeneous broadening will not be considered). Let us consider the case, as previously, of a parallel monochromatic beam propagating in a medium but now the velocity distribution of the atoms is properly accounted for. The probability that the monochromatic beam will interact with a particular atom will depend on the atom's velocity component along the laser beam direction, as well as the frequency of the beam. A total response function for the system can then be obtained by integrating over all the velocity components.

To begin with we write down (3.18) and (3.19) including this velocity dependence. Denoting $n(v)\,dv$ as the number of atoms with velocities between v and $v+dv$, these equations are

$$\frac{dn_2(v)}{dt} = R_2\,\rho(v) - \frac{n_2(v)}{\tau_2} - \left(n_2(v) - \left(\frac{g_2}{g_1}\right)n_1(v)\right)\frac{IB\,f_2(\omega, v)}{c\eta} \tag{3.41}$$

and

$$\frac{dn_1(v)}{dt} = R_1\,\rho(v) - \frac{n_1(v)}{\tau_1} + \frac{n_2(v)}{\tau_{21}} + \left[n_2(v) - \left(\frac{g_2}{g_1}\right)n_1(v)\right]\frac{IB\,f_2(\omega, v)}{c\eta} \tag{3.42}$$

where $\rho(v)$ for a gas is a Maxwell† distribution representing a normalized velocity distribution for the atoms which are formed at some effective temperature T. Thus, atoms are created in the levels E_1 and E_2 at rates R_1 and R_2 with $\rho(v)\,dv$ being the fraction produced with velocity components (along the beam direction) between v and $v+dv$. In the case of a gas $\rho(v)$ is just given by Equation (2.178) divided by $n_0\,dv$. Unless the method of populating the excited states is velocity dependent, this distribution reflects the thermal distribution of atoms in the ground state. (For example, in an optically pumped laser where the ground state atoms are directly transferred to the excited states by photon absorption). f_2 is a normalized Lorenzian with the frequency Doppler shifted to $\omega(1-v/c)$. This is given by the equation

$$f_2(\omega, v) = \frac{\gamma_t/\pi}{(\omega_{12} - \omega(1-v/c))^2 + \gamma_t^2} \tag{3.43}$$

The algebra is now essentially the same as the previous section. At equilibrium the rates dn_1/dt and dn_2/dt are zero and the two equations (3.41) and (3.42) are solved to find n_1 and n_2, assuming that R_1 and R_2 can be held constant. In the case of an isolated two level system n_2 will be given by Equation (2.188) but now written in terms of the A and B coefficients‡. The inversion density from (3.41) and

† Gaussian.

‡ The reader should check this correspondence since it provides a useful exercise.

(3.42) is

$$n_2(v)-(g_2/g_1)n_1(v) = \frac{\rho(v)[R_2\tau_2(1-(g_2/g_1)(\tau_1/\tau_{21}))-R_1(g_2/g_1)\tau_1]}{1+(f_2(\omega,v)IB/\eta c)((g_2/g_1)\tau_1+\tau_2-(g_2/g_1)(\tau_1\tau_2/\tau_{21}))} \tag{3.44}$$

When the intensity is small the second term in the denominator can be ignored and the equation can be directly integrated over all velocity components. Thus, making use of the fact that $\rho(v)$ is normalized we obtain

$$\int_{-\infty}^{+\infty}(n_2(v)-(g_2/g_1)n_1(v))\,dv = n_2^0-(g_2/g_1)n_1^0) \tag{3.45}$$

$$= R_2\tau_2(1-(g_2/g_1)(\tau_1/\tau_{21})) - R_1(g_2/g_1)\tau_1 \tag{3.46}$$

Equation (3.44) is then simplified in the same manner as Section 3.2 to give

$$n_2(v)-(g_2/g_1)n_1(v) = \frac{(n_2^0-(g_2/g_1)n_1^0)\rho(v)}{1+(I_\omega/I_s)\bar{f}_2(\omega,v)} \tag{3.47}$$

where the saturation intensity is given by (3.33) and $\bar{f}_2(\omega,v)$ is the Lorenzian

$$\bar{f}_2(\omega,v) = \frac{f_2(\omega,v)}{f_2(\omega_{12}=\omega(1-v/c))} = \frac{\gamma_t^2}{(\omega_{12}-\omega(1-v/c))^2+\gamma_t^2} \tag{3.48}$$

Figure 3.4 shows how the inversion density varies as a function of the velocity component of the atoms in a gas. The value at low intensities is exactly the same as for a homogeneously broadened transition. As the intensity is increased a dip appears in the distribution when the velocity is equal to $c(1-\omega_{12}/\omega)$. This is just the point at which Equation (3.48) has a maximum value and therefore (3.47) has a minimum value. The beam of radiation of angular frequency ω is said to 'burn a hole' in the velocity distribution. The width of the hole, when the homogeneous broadening is much smaller than the Doppler broadening, is

$$\Gamma_h(\mathrm{v}) = (c/\omega)\gamma_t[1+I_\omega/I_s]^{1/2} \tag{3.49}$$

where Γ_h is the half width at half the hole depth. Notice that (3.47) refers to a single unidirectional beam which is not the case for a laser cavity. Here the light beam is reflected between parallel mirrors so that in general there will be two holes in the distribution at values of velocity $\pm c(1-\omega_{12}/\omega)$. This will be considered later in the chapter when discussing the Lamb dip.

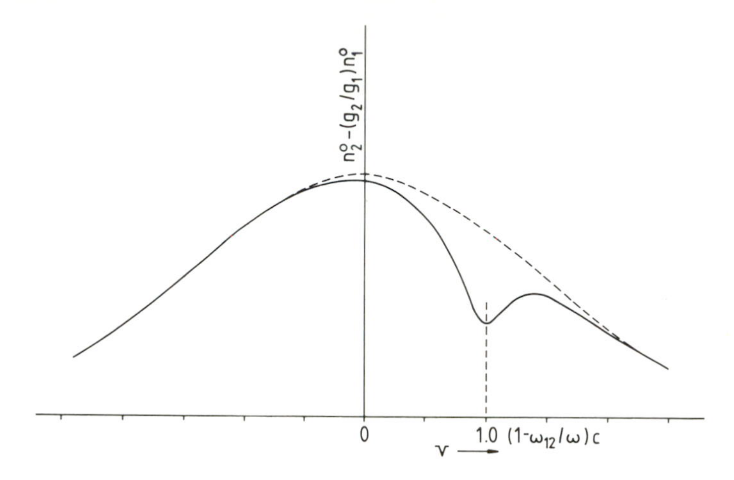

Figure 3.4. Inversion density in units of $n_2^0-(g_2/g_1)n_1^0$ plotted as a function of the atomic velocity, in units of $(1-\omega_{12}/\omega)c$, for an inhomogeneously broadened transition. The homogeneously broadened width is 0·12 times the Doppler width. The dotted line is for low intensity whilst the solid line is when $I_\omega = I_s$.

The gain coefficient analogous to the homogeneous value (3.27) is

$$\alpha(\omega, v) = \frac{(n_2^0-(g_2/g_1)n_1^0)\rho(v)\sigma(\omega, v)}{1+(I_\omega/I_s)\bar{f}_2(\omega, v)} \tag{3.50}$$

where

$$\sigma(\omega, v) = B\hbar\omega f_2(\omega, v)/\eta c = \sigma(\omega_{12})\bar{f}_2(\omega, v) \tag{3.51}$$

Note that the initial inversion density can be written in terms of the lifetime and rates via Equation (3.45) and (3.46). The total response of all the atoms to a particular frequency can now be obtained by integration over all the velocity components. Generally speaking this integral has no direct analytical solution†, but when the homogeneous broadening is much smaller than the Doppler broadening the function $\rho(v)$ can be assumed to be constant for values of v where f_2 makes a significant contribution to the integral. $\rho(v)$ can therefore be removed outside the integral and the inhomogeneous gain coefficient is

$$\alpha(\omega) = (n_2^0-(g_2/g_1)n_1^0)\sigma(\omega_{12})\int_{-\infty}^{+\infty} \frac{\rho(v)\,dv}{1/\bar{f}_2(\omega, v)+I_\omega/I_s} \tag{3.52}$$

which can be solved, when $\rho(v)$ is removed, using the standard definite integral

† These are Voigt integrals discussed in Chapter 2.

$$I = \int_{-\infty}^{+\infty} \frac{dz}{z^2 + k} = \frac{\pi}{(k)^{1/2}} \tag{3.53}$$

to give

$$\alpha_I(\omega) = \frac{(n_2^0) - (g_2/g_1)n_1^0)\pi c \gamma_t \sigma(\omega_{12})\rho(v_r)}{\omega(1 + I_\omega/I_s)^{1/2}} \tag{3.54}$$

where $\rho(v_r)$ is the normalized Gaussian (2.178) at $v_r = c(1 - \omega_{12}/\omega)$. This can be written

$$\begin{aligned}\rho(v_r) &= \left(\frac{m}{2\pi k_b T}\right)^{1/2} \exp\left[\frac{-(\omega - \omega_{12})^2 mc^2}{2\omega_{12}^2 k_b T}\right] \\ &= \frac{1}{v_0\sqrt{\pi}} \exp\left[\frac{-(\omega - \omega_{12})^2 c^2}{\omega_{12}^2 v_0^2}\right]\end{aligned} \tag{3.55}$$

with v_0 being the velocity at which the Gaussian distribution has fallen to 1/e of its value at zero velocity. The factor $\sqrt{\pi}$ ensures the correct normalization when integrating over all velocity components.

To compare the homogeneous and inhomogeneous gain coefficients Equation (3.54) is best written in terms of the Gaussian, $g(\omega)$, formulated in terms of the half width of the distribution. This is of course related to the velocity, v_0, in Equation (3.55) via (2.183). Also, $\sigma(\omega_{12})$ is just $(B\hbar\omega/c\pi\gamma_t)$, so (3.54) is finally reduced to

$$\alpha_I(\omega) = \frac{(n_2^0 - (g_2/g_1)n_1^0)B\hbar\omega\, g(\omega)}{\eta c(1 + I_\omega/I_s)^{1/2}} \tag{3.56}$$

with

$$g(\omega) = \frac{1}{\gamma_I}\left(\frac{\ln 2}{\pi}\right)^{1/2} \exp\left[\frac{-(\omega - \omega_{12})^2 \ln 2}{\gamma_I^2}\right] \tag{3.57}$$

Here, γ_I is the half width of the distribution at half the maximum height. Equation (3.56) can be compared with the homogeneous gain coefficient which can be written, from (3.37), (3.30) and (3.26), as

$$\alpha_h(\omega) = \frac{(n_2^0 - (g_2/g_1)n_1^0)B\hbar\omega\, f(\omega)}{\eta c(1 + (I_\omega/I_s)\bar{f}(\omega))} \tag{3.58}$$

where $f(\omega)$ and $\bar{f}(\omega)$ are given in terms of the half width γ_t by Equations (3.4) and (3.34). The small signal value is therefore essentially the same for both except that for the inhomogeneous case the Lorenzian is replaced by a Gaussian. Our original assumption to evaluate the integral (3.52) meant that the width of the Gaussian was much larger, so for this condition, the small signal gain, which is inversely proportional to the width (γ_I or γ_t for the Lorenzian) will be correspondingly less. (The numerical factors are not much different!) A less obvious result of all this algebra is the different variation of the gain due to saturation. The square root in the denominator of (3.56) arises because the width of the hole in velocity distribution increases with the intensity (as given by (3.49)) at the same time as the depth increases, whereas in the homogeneous case all the atoms are uniformly affected. When the width of the hole increases more atoms are being 'drawn into' the amplification, thus reducing the rate at which saturation takes place.

3.4. Line narrowing and amplified spontaneous emission

The previous analysis considered the change in intensity of a parallel monochromatic beam (from a laser) as it passed through a medium. Here the stimulated emission increases the intensity as long as there is a population inversion, whilst the spontaneous emission is lost. This is the essential basis of a laser amplifier. It consists of a long container of the medium with some form of external pumping to produce the inversion (Figure 3.5). For example the medium

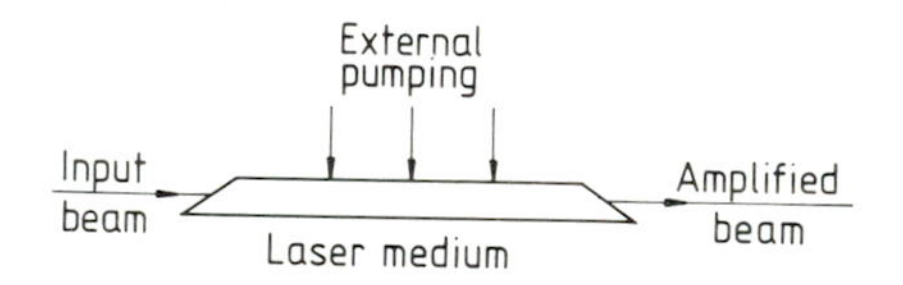

Figure 3.5. Schematic diagram of a laser amplifier.

could be a fluorescent dye and the external pumping could be an intense light source (very often another laser). Such a system will provide amplification for frequencies reasonably close to the line centre of the gain profile of the lasing medium. Providing that the rates R_2 and R_1 can be calculated, and the various constants are known, then the gain coefficient for CW operation can be calculated from the previous equations. However, many amplifiers are for short pulse operation where, at any point along the beam dn_1/dt and dn_2/dt are not zero. These will be considered in subsequent chapters.

Now consider what happens in an amplifier system when the input beam has a broader spectral distribution, for example, that represented by the homogeneous broadening of the transition. When the input and amplified beams are much less than the saturation intensity then the gain coefficient will be given by Equation (3.56) (or 3.58 for homogeneous broadening), with $I/I_s = 0$. The

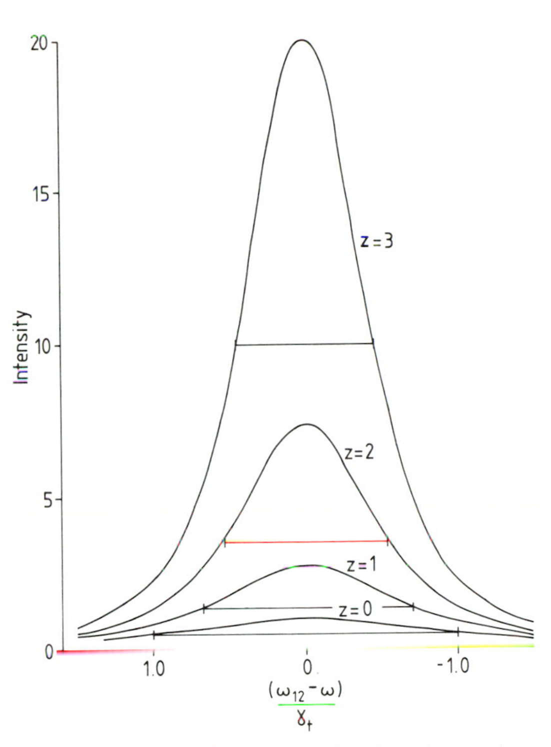

Figure 3.6. Line narrowing when propagating through an active medium when $\alpha(\omega_{12}).z = 0, 1, 2$ and 3. The input beam has the same Lorenzian distribution as the medium.

intensity of a beam at frequency ω increases exponentially as a function of z, according to

$$I(z) = I_0 \exp(\alpha_0(\omega)z) \tag{3.59}$$

where $\alpha_0(\omega)$ is the small signal gain coefficient. This means that the frequencies close to the line centre will be amplified more than those in the wings of the distribution, giving rise to a spectral narrowing of the transmitted beam. This is shown in Figure 3.6. If z is increased indefinitely then at some point saturation will set in, and the narrowing will cease. One reason for this is the change from an exponential variation of intensity to a linear one. For inhomogeneous transitions the line width will actually increase because the frequencies close to ω_{12} burn a hole in the centre of the velocity distribution of the inversion density, leaving the wings unaffected. Frequencies away from line centre can therefore continue to increase exponentially whilst those in the centre increase at a much slower rate.

Next consider how the system shown in Figure 3.5 behaves in the absence of an injected beam. If the pumping is very intense it is possible for a 'partially' coherent beam of radiation to be produced. This is known as Amplified Spontaneous Emission (ASE) and as its name suggests is caused by the amplification of spontaneously emitted light[†]. Although this is emitted in all directions, only along the length of the container will there be sufficient gain for the amplified beam to extract energy by stimulated emission. Since the gain might be large over a fairly large solid angle, it might be expected that the beam would have a large divergence in comparison to a conventional oscillator. In addition, the spectral distribution will be broader than in a normal laser, although it is expected that gain narrowing will produce a beam with a width significantly smaller than the total broadening of the spontaneous emission light. Both the wide divergence and comparatively broad spectral distribution mean that the spatial and temporal coherence will be poor. The temporal coherence is also affected by the fact that the total beam will be made up of many 'components' with different phases, each one initiated by a random photon in the correct direction.

We can put some of these considerations on a mathematical basis in the following manner. The differential equation determining the spectral intensity of light travelling in the positive z direction within the solid angle $d\Omega$ is just

$$\frac{\partial}{\partial z}(I_+(\omega, z)\, d\omega) = \alpha(\omega) I_+(\omega, z)\, d\omega + \hbar\omega A_{21}\, n_2\, f(\omega)\, d\omega \frac{d\Omega}{4\pi} \tag{3.60}$$

Here α is the gain coefficient and $f(\omega)$ is the general line broadened shape. This equation is obtained by considering the change in intensity across a short slab of length dz, in the same manner as Equations (3.5) to (3.8). Now, however, we include the spontaneous emission via the second part of the equation. This can normally be neglected in a laser cavity because $d\Omega$ is vanishingly small. Notice that the equation makes no reference to any coherence because it only involves intensities not amplitudes. A similar equation to this will apply for the beam travelling in the opposite direction. In general, these two equations are not independent because $\alpha(\omega)$ will depend on I_+ and I_- through the saturation term. This makes the equations complicated, particularly if the medium has a more general shape. To simplify matters, saturation is ignored by putting $\alpha = \alpha_0$ and the intensity at distance, z, subject to the boundary condition $I_+(z=0) = 0$, is

$$I_+(\omega, z) = \frac{A_{21}\, c n_2^0}{B(n_2^0 - (g_2/g_1) n_1^0)} (\exp(\alpha_0 z) - 1) \frac{d\Omega}{4\pi} \tag{3.61}$$

$$= \frac{\hbar\omega^3 n_2^0}{\pi^2 c^2 (n_2^0 - (g_2/g_1) n_1^0)} (\exp(\alpha_0 z) - 1) \frac{d\Omega}{4\pi} \tag{3.62}$$

where Equation (2.24) has been used for the ratio (A_{21}/B_{21}).

To show that this equation is sensible, the two extremes can be examined.

† This is also true in a sense of a laser cavity described in the next section. For a laser, a single photon in the correct direction switches on the oscillations.

For an optically thin sample ($\alpha_0 z \ll 1$) the exponential can be approximated to $(1+\alpha_0 z)$ and the intensity is

$$I_+(\omega, z) = A_{21}\, n_2^0\, \hbar\omega z\, f(\omega)(d\Omega/4\pi) \tag{3.63}$$

which is just the amount of light spontaneously emitted into the solid angle $d\Omega$. At the other extreme ($a^{\alpha_0 z} \gg 1$) the intensity varies in an exponential manner of an unsaturated amplifier, with the constant I_0 given by the first part of (3.61). When a system is close to producing amplified spontaneous emission it is generally unsuitable for an amplifier because energy may be extracted before the laser pulse arrives. Energy will be transferred (in both directions) to this broad-band emission rather than into the injected beam. A limit is therefore put on the gain of any laser amplifier. In practice amplifiers are often run at low gain, and where a large total power is required the beam size perpendicular to the direction of propagation is increased. This increases the stored energy in the amplifier without altering its gain characteristics. Of course the diameter cannot be increased indefinitely because the amplifier may produce ASE in this direction! Also the beam area must be matched to the diameter of the amplifier.

ASE should be carefully distinguished from super-radiance. In this latter process the oscillating dipoles are all in phase and can collectively emit a coherent pulse of radiation. The whole system therefore acts like a single macroscopic dipole. The process requires some form of coherent pumping and it cannot be analyzed using our simple intensity method.

3.5. Anomalous dispersion

A wave travelling through the laser medium will undergo the effects of dispersion, by which it is meant that its phase velocity (c/η) will be dependent on its frequency, ω. In Section 2.9 it was shown that the refractive index of a gas undergoes a rapid change (Figure 2.9) at a resonant frequency ω_{12}. This anomalous dispersion has to be carefully considered in any description of a laser. For example, a short pulse of light will alter its amplitude-frequency profile as it propagates in the medium. Also the modes in a standing wave laser will be shifted in frequency, dependent on their position relative to ω_{12} (mode pulling). For a gas at low pressure the refractive index can be obtained from Equation (2.167) by merely replacing $K(\omega)$ with $-\alpha(\omega)$. This gives

$$\eta(\omega)-1 = -\alpha(\omega)c(\omega_{12}-\omega)/2\gamma_t\,\omega_{12} \tag{3.64}$$

where $2\gamma_t$ is the full width of the distribution at half maximum. The refractive index therefore shows the reverse effect to that plotted in Figure 2.9. For frequencies below ω_{12} the index is less than one whilst for frequencies above ω_{12} it is greater than unity.

In a solid or liquid dielectric the equations are not so simple because the bulk polarization of the medium has to be accounted for. In writing down (2.164) and (2.165) it was assumed that the electric field at the atomic site was the same as that in the oscillating electromagnetic field. When the polarization is included the equations for the refractive index are more complicated. Nevertheless, they still show similar anomalous dispersion at the resonant frequency. Now, however, there is a small shift in the resonant frequency away from the line centre ω_{12}.

3.6. The laser cavity

It is now time to consider how the previous analysis in this chapter can be applied to a real laser cavity. Figure 3.7 shows a schematic diagram of a standing-wave laser which comprise the majority of lasers. The active medium, in which the

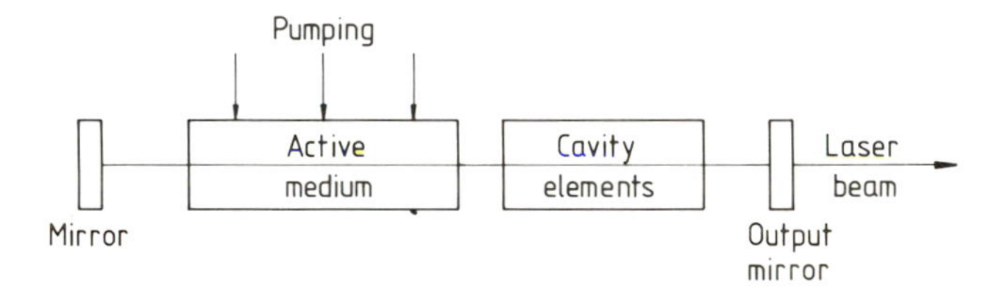

Figure 3.7. Schematic diagram of a standing wave laser cavity.

population inversion is induced, is placed between two parallel mirrors and the light is successively amplified as it passes through it. The output mirror (or coupler) is constructed in such a way as to allow a small fraction of the incident light to escape from the cavity. Extra elements, such as Q-switches or etalons, are often included in the cavity to alter the characteristics of the output light. The method of producing the inversion depends on the type of laser. For example, in a gas laser the pumping is often effected by an electrical arc discharge. In a doped insulator laser (e.g. ruby) the inversion is produced by optical pumping using a flash tube. Whatever the method, the aim is to convert energy from the pumping source into laser light. At first sight it may seem that the only purpose of the cavity is to provide an effective increase in the total gain without having an unduly long laser system. This is of course true, but the influence of the cavity is more subtle. The repeated passage of the beam sets up a system of standing waves which allow energy to be extracted only over a narrow frequency range, much smaller than the frequency profile of the gain coefficient, α. Indeed, the linewidth of the output light can often be much narrower than the natural broadening of the transition! In many ways the system can be likened to a high Q electric oscillator circuit. The amplifier in this is equivalent to the active medium whilst the components of the electrical filter (coil and capacitor) are equivalent to the mirrors and other intracavity elements. For the electrical case only frequencies close to the natural frequency of the external circuit are amplified. Energy is thus 'funnelled' into a narrow frequency range. By analogy therefore the open loop gain of the laser oscillator can be given by Equation (3.31) with z equal to the length of the laser medium. Notice however that this refers to intensity only and a correct analysis would require that we deal with amplitudes. In this way the phase changes during each passage through the medium and reflection at the mirrors could be properly accounted for. The frequency response of the system is determined by the homogeneous or inhomogeneous broadening. A gas laser therefore has a fairly narrow range of frequencies over which it can be made to 'oscillate'. On the other hand, the homogeneous broadening in a dye laser is very large, and these types of laser can be tuned over a much broader range of frequencies. Subsequent chapters will deal specifically with different types of lasers.

3.7. *Threshold condition for laser oscillation: Q value*

Let us now consider the basic conditions in the laser cavity which determine the threshold for laser oscillation. In the first instance, it is convenient to get a simple picture by analysing the system in terms of the intensity of the electromagnetic wave. Furthermore, the propagating wave will be considered to have an infinite extent orthogonal to the direction of propagation (hypothetical of course!). In this way the effects of diffraction are ignored. Consider the scheme, shown in Figure 3.8, where the beam is imagined to start after leaving the first mirror, M1. It then passes through the active medium and, if there are no losses, or saturation, its amplitude will increase by the factor $\exp(\alpha L)$, where α is the small signal gain coefficient. To account for losses such as scattering or absorption a coefficient β is introduced. β represents the fractional change in intensity per unit thickness for a very thin section of the medium. The gradient of the intensity is therefore

$$\frac{dI}{dz} = (\alpha - \beta)I \tag{3.65}$$

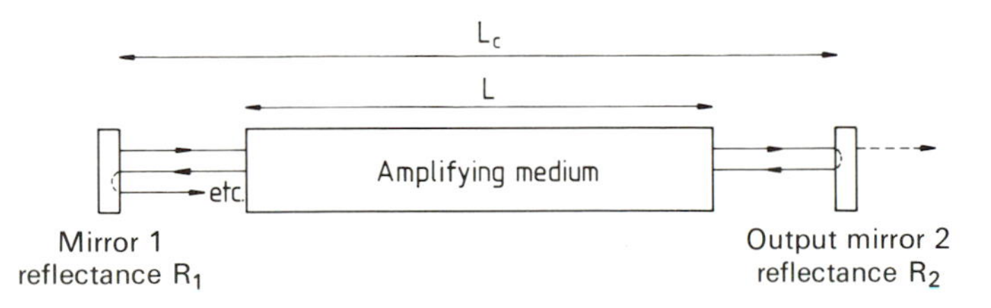

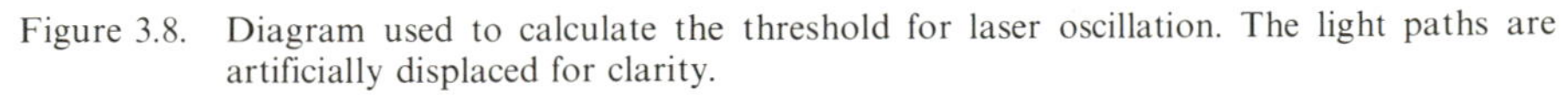

Figure 3.8. Diagram used to calculate the threshold for laser oscillation. The light paths are artificially displaced for clarity.

and the intensity after travelling a distance L is

$$I(L) = I_0 \exp((\alpha - \beta)L) \tag{3.66}$$

On striking the second mirror, M2, a fraction R_2 is returned to the cavity and $(1-R_2)$ leaves the cavity. The beam is then further amplified by a factor $\exp[(\alpha-\beta)L]$ before it strikes M1 and a fraction $R1$ is returned to the cavity. The total amplification in intensity for the round trip is therefore

$$G_R = R_1 R_2 \exp(2(\alpha - \beta)L) \tag{3.67}$$

For the laser to oscillate this gain must be greater than unity. The threshold value of α in the absence of diffraction losses is therefore

$$\alpha_{\mathrm{TH}} = \beta + \ln(1/R_1 R_2)/2L \tag{3.68}$$

If β, R_1 and R_2 are known then this threshold value can be compared with the small signal value (3.29) to see if oscillations are possible. Once the oscillations are established the value of the gain coefficient is reduced because of saturation. The equilibrium value is when the increase in intensity for a round trip is exactly balanced by the total losses in the cavity. Because the variation in intensity through the laser amplifying material may no longer be exponential the value of α need not necessarily be given by (3.68). However, in many cases, α and β are sufficiently small so that very many transits through the medium are required to reach saturation, and in these circumstances the fractional change in intensity for the round trip can be approximated to

$$\frac{I}{I_0} \simeq R_1 R_2(1 + 2(\alpha - \beta)L) \tag{3.69}$$

This must be unity at equilibrium and hence the gain coefficient is

$$\alpha \simeq \beta + \frac{(1 - R_1 R_2)}{2L} \equiv \frac{\delta_c}{2L} \tag{3.70}$$

where δ_c is the total fraction of intensity lost† during the round trip, which may be suitably modified to account for diffraction losses. Notice that the hypothesis of a small change in intensity traversing the medium implies that R_1 and R_2 must be close to one. This is just a mathematical statement of the obvious conclusion that the maximum amount that can be extracted from the output mirror, at equilibrium, is what is put in during one round trip. Unless the inversion density

† The total fraction of absorption and scattering is $2L\beta$.

is extremely high, the output mirror will only have a transmission of a few percent. In many pulsed lasers (e.g. flashlamp pumped system), the initial inversion density may be very high, and the output mirrors often have transmissions of 20 or 30%. For some of these cases the inversion density can only be maintained for a short period before saturation causes the laser to cease oscillating. For a CW laser the output intensity at equilibrium can be obtained by equating the distributed cavity loss (3.70) to the gain equation (3.27). Since the gain is frequency dependent it is necessary to know the output frequencies of the laser. These are considered in the next section. When more than one frequency (cavity mode) can oscillate the output depends on whether the transition is homogeneously or inhomogeneously broadened. This is also discussed in the next section.

The losses in the laser cavity essentially determine the Q value of the oscillator. This can be defined by the equation

$$Q = \omega \frac{\text{Energy stored in the cavity}}{\text{Energy lost per second}} \tag{3.71}$$

When the cavity losses are very low, the Q value is high, and the damping in the 'circuit' is small. A high Q value is also synonymous with a narrow linewidth and for all resonant oscillators, we have

$$Q = \frac{\omega_{12}}{\Delta\omega_c} \tag{3.72}$$

where ω_{12} is the resonant frequency and $\Delta\omega_c$ is the full width at half height of the resonant mode of the passive cavity. Note carefully that this is not the laser output linewidth although the two are related.

Q can be related to the cavity parameters by imagining that the laser excitation is suddenly extinguished. Equation (3.71) can then be written

$$Q = \frac{\omega W}{-(dW/dt)} \tag{3.73}$$

which has a solution

$$W = W_0 \exp(-t\omega/Q)$$

where W_0 is the energy at time zero. The energy thus decreases in amplitude with a characteristic time constant, t_0, given by

$$t_0 = \frac{Q}{\omega} = \frac{1}{\Delta\omega_c} \tag{3.74}$$

This time constant can be obtained by considering the loss of intensity in the cavity after switch off time. In one round trip there is a total fractional loss δ_c and this takes place in a time $c/2L_c$. For damping over many cycles the intensity is determined by the equation

$$\frac{dI}{dt} = -\frac{I_c \delta_c}{2L_c} \tag{3.75}$$

The time constant for the decay of intensity must be the same as that for energy and therefore

$$\frac{1}{t_0} = \Delta\omega_c = \frac{c\delta_c}{2L_c} \tag{3.76}$$

The cavity losses therefore determine the Q value which can now be related to the gain coefficients via Equations (3.70) and (3.68). An estimate of the Q value for a CW laser operating in the visible spectrum can be obtained from (3.70). Take, for example, a He–Ne laser (λ = 533 nm) with a 1 m cavity and assume, for simplicity, that the scattering, absorption and diffraction losses are much smaller than the output losses. A typical laser might have $1-R_1 R_2 = 0{\cdot}02$, giving a Q value of the order 10^9. This is considerably larger than a good microwave cavity, where Q might be as high as 10^4.

3.8. *Longitudinal cavity modes*

The electric field in a laser cavity forms a standing wave pattern in an analogous way to the standing vibrational waves in a stretched string which is rigidly held at both ends. Only certain frequencies, determined by the cavity geometry, can resonate and extract energy from the amplifying medium. In a stretched string the analysis is relatively simple because there is only one spatial dimension (length along the string) and the boundary conditions are that there should be zero amplitude at the ends. For a laser cavity the problem is not nearly so simple because the system is three dimensional with extra boundary conditions besides those imposed by the mirrors. In addition, the active medium may not extend the whole length of the cavity so, in principle, different conditions apply at different points along the cavity. Generally the analysis is very complicated and solutions can only be obtained for geometries which have at least one symmetry axis. Even in these cases numerical techniques must be used.

To begin with, we make an analysis of the allowed longitudinal modes in a passive cavity and defer a more detailed study of the mode structure until the next section. The longitudinal modes are found by an identical analysis to that used in a Fabry–Pérot etalon. In terms of our simple view of the laser the mirrors

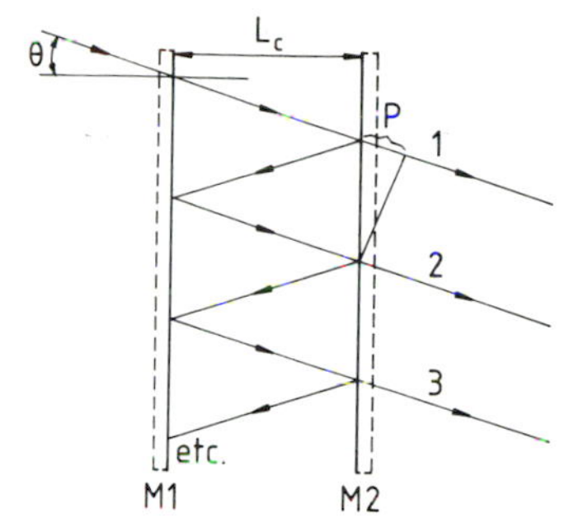

Figure 3.9. Successive reflections in a Fabry–Pérot etalon with plane mirrors.

therefore have a dual purpose. Firstly, they provide the essential feedback for sustained oscillation and secondly, they act as a narrow bandpass filter. The total frequency response of the system will therefore be a combination of that due to amplifying medium and that imposed by the cavity geometry.

The relevant diagram for calculating the transmission of a Fabry–Pérot etalon is shown in Figure 3.9. M1 and M2 are plane reflecting mirrors and L_c is their separation. A plane wave incident at an angle θ undergoes multiple reflections inside the cavity. At each mirror a certain fraction of the light is reflected and the remainder is transmitted (assuming no absorption). To calculate the transmitted intensity the total amplitude represented by the successive rays 1, 2, 3, etc is first calculated. This analysis therefore requires a knowledge of the relevant phases, unlike our simple analysis in Section 3.7. The calculation is simplified if the incident driving wave is linearly polarized with the electric field vector either perpendicular or parallel to the plane of the diagram. It is then a relatively simple matter to generalize to any state of polarization. In fact, for most lasers, the Brewster windows, used to confine active medium, means that amplification is often only possible for plane polarized light. The electric field vector of the amplified wave then lies in the plane defined by the Brewster angle.

Let the ratio of the electric field on reflection and transmission be defined by the symbols ρ and σ. The electric field amplitude in the wave labelled 1 will therefore be

$$E_1 = E_0 \sigma_1 \sigma_2 \tag{3.77}$$

where E_0 is the external driving amplitude. Similarly the electric fields in the plane wave 2, 3, 4 etc will have amplitudes

$$\begin{aligned} E_2 &= E_0 \sigma_1 \sigma_2 \rho_1 \rho_2 \\ E_3 &= E_0 \sigma_1 \sigma_2 \rho_1^2 \rho_2^2 \\ E_4 &= E_0 \sigma_1 \sigma_2 \rho_1^3 \rho_2^3 \\ &\ldots\ldots\ldots\ldots\ldots \end{aligned} \tag{3.78}$$

Now these cannot be summed directly to get the total amplitude, since their phases may not necessarily be the same. To calculate the phase change between successive terms, we first note that the path difference between any two consecutive wave fronts is

$$\Delta s = (2L_c/\cos\theta) - P = 2L_c\cos\theta \tag{3.79}$$

This path difference corresponds to a phase change ($\eta = 1$)

$$\phi = 2\pi(2L_c\cos\theta/\lambda) = (\omega/c)2L_c\cos\theta \tag{3.80}$$

and so the total field amplitude of the transmitted wave is

$$E_T = E_0\sigma_1\sigma_2(1+\rho_1\rho_2\exp(-i\phi)+\rho_1^2\rho_2^2\exp(-i2\phi)+\rho_1^3\rho_2^3\exp(-i3\phi)+\ldots) \tag{3.81}$$

With the aid of the binomial expansion ($x^2<1$)

$$\frac{1}{(1-x)} = 1+x+x^2+x^3+\ldots \tag{3.82}$$

Equation (3.81) becomes

$$E_T = \frac{E_0\sigma_1\sigma_2}{(1-\rho_1\rho_2\exp(-i\phi))} \tag{3.83}$$

Using the expression (2.99), the transmission is

$$T = \frac{I_T}{I_0} = \frac{\sigma_1^2\sigma_2^2}{1+\rho_1^2\rho_2^2-2\rho_1\rho_2\cos\phi} \tag{3.84}$$

which can be conveniently written in terms of the reflection coefficients, R_1 and R_2, as

$$T = \frac{(1-R_1)(1-R_2)}{(1-(R_1R_2)^{1/2})^2+4(R_1R_2)^{1/2}\sin^2(\phi/2)} \tag{3.85}$$

Here the reflection coefficients are given in terms of ρ and σ by the equations

$$R_1 = \rho_1^2 \tag{3.86}$$

$$R_2 = \rho_2^2 \tag{3.87}$$

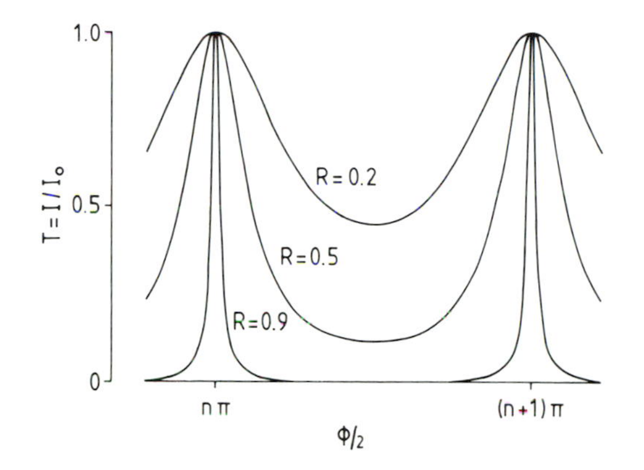

Figure 3.10 Transmission of a Fabry–Pérot etalon when $R_1 = R_2 = R$, for $R = 0{\cdot}9$, $0{\cdot}5$ and $0{\cdot}2$, plotted as a function of the phase angle ϕ. The frequency shift corresponding to $\Delta(\phi/2) = \pi$ is known as the free spectral range and is c/L_c. This is also the mode spacing of an equivalent cavity.

$$1 - R_1 = \sigma_1^2 \tag{3.88}$$

and

$$1 - R_2 = \sigma_2^2 \tag{3.89}$$

The transmitted intensity represented by Equation (3.85) is plotted in Figure 3.10. It consists of a series of peaks separated in frequency space by the interval

$$\Delta\nu = \frac{\Delta\omega}{2\pi} = \left(\frac{c}{2L_c}\right)\cos\theta \tag{3.90}$$

These peaks correspond to the condition that the phase shifts of successive waves are an integral number times 2π.

To compare the transmission of an externally driven etalon with the frequency response of a laser cavity we need to put $\theta = 0$ in the previous equations. The quantity given by (3.90) is the free spectral range (FSR), and in the laser it is the cavity mode spacing. (Notice that Equation (3.85) cannot be applied to a laser cavity as it stands because the fraction $(1 - R_1)$ must be greater than zero to allow light into the cavity. In a laser the light is generated inside so this can be ignored.) For both cases the spiked frequency response arises because a system of standing waves are set up between the mirrors. In the etalon the energy in these modes builds up until the output intensity matches the input intensity, whilst, in the cavity, the output intensity must match the fraction of energy fed into coherent emission. The longitudinal standing wave pattern in a laser cavity gives the condition

$$\frac{q\lambda}{2} = L_c \tag{3.91}$$

in the same way as the case of vibrations of a stretched string. Accordingly the angular frequencies of the modes are

$$\omega_q = \frac{q\pi c}{L_c} \tag{3.92}$$

where q is an integer defining the longitudinal mode number (or order). This is exactly the same as the analysis in Section 2.1 for deriving the allowed modes in a rectangular box, but here we are restricting our attention to a single dimension. Each different frequency which is amplified (or transmitted) corresponds to a different number q, and the separation of any two frequencies is therefore given by Equation (3.90) with $\theta = 0$. A high Q corresponds to a large amount of stored energy in the modes which arises when the damping or cavity losses are small. The Q value of the etalon also determines the width of the transmitted peaks. Figure 3.10 shows that this frequency spread depends on the losses at the mirrors, a sharp line being synonymous with a high coefficient of reflection at both mirrors. This linewidth can be obtained from Equation (3.85) by calculating the point at which the transmission is reduced by one half. Since the maximum value is when $\sin\phi = 0$, the half height frequencies are solutions of the equation

$$4(R_1 R_2)^{1/2} \sin^2(\omega L_c/c) = (1-(R_1 R_2)^{1/2})^2 \tag{3.93}$$

For low losses (high R_1 and R_2) the half points are close to values where $\sin\theta = 0$, so the approximation, $\sin\theta = \theta$ can be used, and

$$\omega_{1/2} = \omega_{12} \pm \frac{c(1-(R_1 R_2)^{1/2})}{2L_c(R_1 R_2)^{1/4}} \tag{3.94}$$

The half height width is therefore

$$\Delta\omega_{1/2} = \frac{c(1-(R_1 R_2)^{1/2})}{L_c(R_1 R_2)^{1/4}} \tag{3.95}$$

and the Q value is

$$Q = \frac{\omega_{12}}{\Delta\omega} = \frac{\omega_{12} L_c(R_1 R_2)^{1/4}}{c(1-(R_1 R_2)^{1/2})} \tag{3.96}$$

This can be compared with our previous analysis, given by Equations (3.70) and (3.76). The two agree only when R_1 and R_2 are close to unity, which was our original assumption in Section 3.7.

To close this Section, Figure 3.11 shows the small signal gain of an oscillator when the transition is inhomogeneously broadened. This consists of a series of

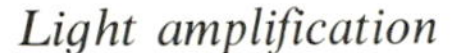

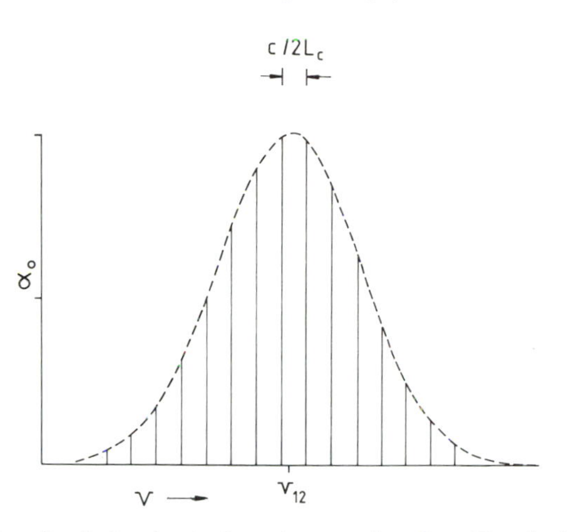

Figure 3.11. Small signal gain for a laser cavity where Doppler broadening is dominant. The cavity length is such that the mode spacing is much smaller than the width of the Gaussian.

narrow peaks which span the Doppler lineshape. Each peak is separated from its neighbour by the mode spacing $c/2L_c$†. If the transition is homogeneously broadened then the Gaussian envelope would be a Lorenzian. However, because of saturation effects, the output of a homogeneously broadened laser consists of only one mode. This effect is discussed more fully in Section 3.13.

3.9. *Transverse mode structure*

The previous section gave an analysis for the longitudinal standing waves and ignored any variation of the electric field normal to the laser cavity length, i.e. the assumption was for a hypothetical plane wave. For a correct analysis the full three dimensional wave equation would need to be solved for the boundary conditions imposed by the laser geometry. Unfortunately, it is not possible to find completely general solutions although for certain approximations the transverse mode structure can be found. In particular, for a linearly polarized transverse electromagnetic (TEM) wave, where the wavelength is small compared to the cavity dimensions, then the transverse mode structure‡ can be found by repeated application of Huygens' principle.

Consider a passive cavity consisting of two square flat mirrors separated by a distance L_c, as shown in Figure 3.12. The transverse electric field on M2 is related to the electric field on M1 by the Fresnel–Kirchoff integral.

† In a medium of refractive index η, this is $c/(\eta 2L_c)$.

‡ Following the original article by A.G. Fox and Tingye Li in *Bell System Tech. J.* **40** (1961) p. 453.

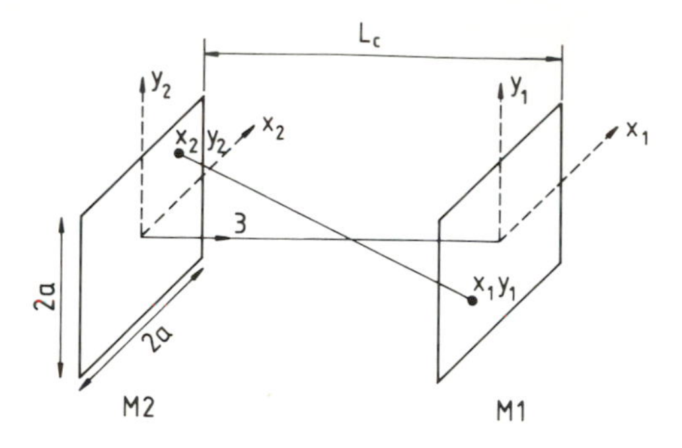

Figure 3.12. Geometry for calculating the electric field distribution on rectangular mirrors.

$$E_2(x_2, y_2) = (ik/4\pi)\iint_s E_1(x_1, y_1)(\exp(-ikr)/r)(1+\cos\theta)\,dx_1\,dy_1 \quad (3.97)$$

where the integral is taken over the surface of M1. Here r is the distance between the point x_1, y_1 on M1 and x_2, y_2 on M2. For the square geometry in the figure the field can be separated into the product of two parts, each dependent only on the single variable x or y. Thus (3.97) becomes

$$E_2^x(x_2)E_2^y(y_2) = (ik/4\pi)\int_{-a}^{+a}\int_{-a}^{+a} E_1^x(x_1)E_1^y(y_1)(\exp(-ikr)/r)(1+L_c/r)\,dx_1\,dy_1 \quad (3.98)$$

where

$$r = (L_c^2+(x_2-x_1)^2+(y_2-y_1)^2)^{1/2} \quad (3.99)$$

and, when a is much smaller than the cavity length, the integral (3.98) can be approximated to

$$E_2^x(x_2)E_2^y(y_2) = \frac{ik\exp(-ikL_c)}{2\pi L_c}\int_{-a}^{+a}\int_{-a}^{+a} E_1^x(x_1)E_1^y(y_1) \times \exp[-ik((x_2-x_1)^2+(y_2-y_1)^2)/2L_c]\,dx_1\,dy_1 \quad (3.100)$$

Now Equations (3.100) and (3.97) are just prescriptions for calculating the propagation of the electric field from one mirror to the other. The other factor which needs to be decided is the method for describing a stable mode. Consider what happens if a particular wave 'pattern' travels backwards and forwards between the mirrors. If this is a stable mode, then, after a certain large number of oscillations, the distribution on any one mirror becomes fixed, although its amplitude will decrease at each transit. This condition is expressed mathematically by writing the distribution after a further r transits as

$$E_r = (\gamma)^r E \tag{3.101}$$

where E is a stable field pattern which does not vary from reflection to reflection. γ is a complex constant describing the amplitude and phase change per transit. Combining (3.101) and (3.100) the equations determining the stable modes of a cavity with square mirrors are

$$\gamma_m E_m(x_2) = \frac{\exp(i\pi/4)}{(2\pi L_c/k)^{1/2}} \int_{-a}^{a} E_m(x_1) \exp[-i(k/2L_c)(x_2 - x_1)^2] \, dx_1 \tag{3.102}$$

and

$$\gamma_n E_n(y_2) = \frac{\exp(i\pi/4)}{(2\pi L_c/k)^{1/2}} \int_{-a}^{a} E_n(y_1) \exp[-i(k/2L_c)(y_2 - y_1)^2] \, dy_1 \tag{3.103}$$

Each mode in the two directions is labelled with the integers m and n. The distributions E_m and E_n are the eigenfunctions whilst γ_m and γ_n are the eigenvalues. (Some readers will be familiar with eigenfunction equations when the operator contains differentials rather than integrals. For example, the stationary states (stable modes!) of the hydrogen atom are described by the operator equation $H\psi = E\psi$, where H is given by Equation (1.2). If the laser was totally enclosed in a rectangular box with known boundary conditions then the wave equation (2.6) could be used in an entirely analogous manner. This method is used extensively in the theory of waveguides.)

Exact solutions to (3.102), or (3.103), can only be obtained by numerical techniques. To do this, a trial solution is first substituted into the RHS of the equation which is then integrated numerically. The new solution is reinserted into the equation and the process repeated. After many transits a function is found which does not alter its form upon evaluating the integral. The stable electric field distributions are then given by the product

$$E_{mn} = E_m(x) E_n(y) \tag{3.104}$$

and $\gamma_n \gamma_m$ is the constant γ, describing the total change in amplitude and phase for each traversal of the beam. The total power loss due to diffraction is therefore $1 - \gamma\gamma^*$ per transit. Notice that Equation (3.102) is identical in form to (3.103) so that only one needs to be solved. The problem thus corresponds to one of determining the modes in a cavity with 'infinite-strip' mirrors. A complete structure is written $(\mathrm{TEM})_{mnq}$ where q is the longitudinal mode number given by Equation (3.91). m and n define the number of nodes (points where the amplitude crosses the axis) in the x and y directions. For the longitudinal structure there are

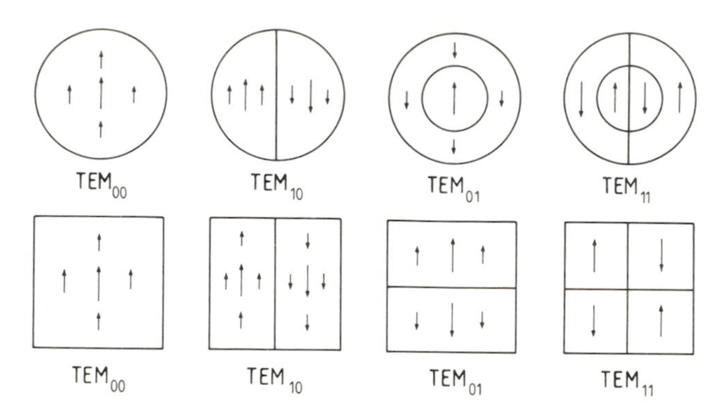

Figure 3.13. Electric field configurations and nodal lines for the four lowest order normal modes. (After A.G. Fox and Tingye Li, *Bell Systems Tech. J.* **40** (1961) 453, © AT&T, 1961.)

$q-1$ nodes along the cavity length, in line with the requirement that an integer half-wavelengths fit into the cavity.

A similar analysis can be carried out for flat circular mirrors. Here the field is also written as a product of two terms, one dependant on the radial position, R, from the centre of the mirror and the other on the azimuthal angle, ϕ. The distributions are now labelled with the indices p, l and q, where p and l are the number of radial and circular modal lines. Figure 3.13 shows the general pattern of TEM fields on square and circular plane mirrors for the lowest stable modes. Another common laser geometry is the confocal resonator shown in Figure 3.14. Intuitively, one would expect that such a system would have lower diffraction losses, and this is borne out by the results of numerical calculations. Figure 3.15 is a plot of the power loss (per transit) as a function of the Fresnel number, F, defined by the equation

$$F = \left(\frac{a^2}{L_c\lambda}\right) \tag{3.105}$$

It is a fairly simple matter to show that the eigenvalues in Equation (3.98) can be parameterized in terms of this single variable, and the same consideration applies for other geometries. Figure 3.15 shows that the lowest diffraction losses, and, incidently, the smallest beam divergence, corresponds to the TEM_{00} mode. In

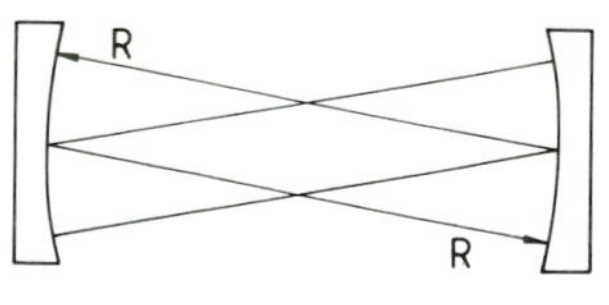

Figure 3.14. The confocal resonator. Both mirrors have the same radius of curvature, R, and are separated by the distance R.

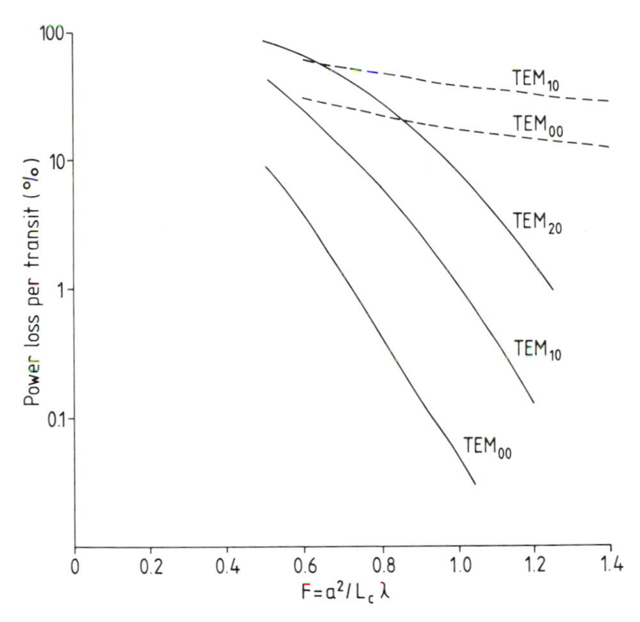

Figure 3.15. The power loss per transit for circular mirror. The dotted line is for a plane parallel cavity whilst the solid line is for a confocal geometry.
(After A.G. Fox and Tingye L, *Bell System Tech. J.* **40** (1961) 453, © AT&T, 1961.)

the confocal geometry the difference between this mode and higher ones is much larger, giving this type of cavity a high degree of mode stability. In terms of practical usage, most lasers are required to operate on the TEM_{00} mode. This is because the beam has a smaller divergence and can be utilized more effectively. In addition the wave front has a common phase unlike the higher order modes. (See Figure 3.13.) A frequent practice is to include constricting circular apertures along the laser cavity to preferentially increase the diffraction losses of unwanted modes. Another virtue of cavities with spherical mirrors† is their insensitivity to mirror alignment. This is in contrast to a resonator with plane mirrors, where, as one might expect, the general stability and diffraction losses are extremely sensitive to the accuracy to which the mirrors can be maintained parallel. Most stable resonators are now designed with at least one mirror which is concave. Not only are these geometries more tolerant to misalignment of the mirrors but they have much lower diffraction losses than the corresponding plane parallel cavity cf. Figure 3.15.

It is possible to obtain approximate analytical solutions for all these cavity geometries. These give very useful insight into the general shapes of the electric field distributions. The technique is to extend the limits of integration of (3.98) to infinity which amounts to saying that the far field diffraction pattern is the same as the near field distribution. It can be seen from the form of the integral that such

† The confocal geometry is a rather special case as far as stability is concerned. See Section 3.11.

a procedure is a good approximation only for large Fresnel numbers. The solutions for the rectangular case have the general form

$$E_{mn}(x, y) = E_0 H_m(\sqrt{2}x/w)H_n(\sqrt{2}y/w)\exp[-(x^2+y^2)/w^2] \qquad (3.106)$$

where H_m and H_n are the Hermite polynomials of order m and n. w is a parameter with the dimensions of length related to the cavity geometry and dimensions. It is often called the spot size, or beam radius, since it defines the point at which the TEM_{00} distribution has fallen to 1/e of its value at $x, y=0$. The functions given by (3.106) are, in fact, the same as the eigenfunctions for the two dimensional harmonic oscillator in quantum mechanics. (The differential equation for a single dimension is given by Equation (1.25) of Chapter 1.) It is obvious that such solutions form a complete set of orthonormal functions for the description of any field distribution. Notice, also, that this approximation to the Fresnel–Kirchoff equation (3.97) reduces to a two-dimensional Fourier transform in the Fraunhofer approximation. The Gaussian functions (and the Hermite polynomials) are the only stable modes because they are their own Fourier transforms.

When the problem is cylindrically symmetric (e.g. circular mirrors and a circular cross section for the amplifying medium) the equation determining the distribution can be written

$$E(R_2, \theta_2) = C\int_0^{\infty}\int_0^{2\pi} E(R_1, \theta)\exp[-ikR_1 R_2\cos(\theta_1-\theta_2)/r]R_1\, dR_1\, d\theta_1 \qquad (3.107)$$

where R and θ are the polar co-ordinates of the positions on the mirrors. The constant C now incorporates the eigenvalue (or loss parameter) γ. In this case the solutions have the form

$$E_{pl}(R, \theta) = E_0(\sqrt{2}R/w)^l L_p^l(2R^2/w^2)\exp(-R^2/w^2)\begin{Bmatrix}\sin(l\theta)\\ \cos(l\theta)\end{Bmatrix} \qquad (3.108)$$

where the sine function is chosen for odd values of l and the cosine for even values. L_p^l is the Laguerre polynomial of order pl. For the confocal cavity the beam spot size, w, is given by the equation

$$w = \left(\frac{L_c\lambda}{\pi}\right)^{1/2} \qquad (3.109)$$

so the TEM distribution corresponds to Gaussian with a half width at half height of

$$w_{1/2} = (\ln 0{\cdot}5)^{1/2}w \qquad (3.110)$$

3.10. Gaussian optics

The previous section leads quite generally into the subject of Gaussian optics which is now briefly described. It has wide applicability for describing the lowest order field patterns within the cavity as well as being useful for understanding the transport and matching of laser beams. Consider first the $(\mathrm{TEM})_{00}$ solutions, (3.108) and (3.106), for the infinite aperture passive cavity. The field distribution along the whole cavity can also be described by similar Gaussian functions with suitably adjusted parameters of phase and beam spot size. In fact, these functions must correspond to simple solutions of the scalar wave equation

$$\nabla^2 u(\boldsymbol{r}) + k^2 u(\boldsymbol{r}) = 0 \tag{3.111}$$

which is found by separating out the spatial and time-dependent part of the full equation (2.57). u can then refer to any time-averaged scalar quantity describing the field, for example, the magnitude of the electric vector. Now we are looking for simple solutions to this equation of the form

$$u(\boldsymbol{r}) = \psi(x, y, z)\exp(-ikz) \tag{3.112}$$

where z corresponds to the direction of propagation of the beam. Here ψ represents the shape of the wavefront or its departure from a plane wave form. Substituting (3.112) into (3.113), and ignoring the term $\partial^2\psi/\partial z^2$ because ψ is relatively slowly varying, gives

$$\frac{\partial^2\psi}{\partial x^2} + \frac{\partial^2\psi}{\partial y^2} = 2ik\frac{\partial\psi}{\partial z} \tag{3.113}$$

This is the same as the two dimensional time dependent Schrödinger equation (z replaces t!). The solution to (3.112) which corresponds to our Gaussian beam can therefore be written in the form

$$u(R, z) = (w_0/w(z))\exp(-r^2/w^2(z))\exp(-i(kz-\phi))\exp(-ikr^2/2R(z)) \tag{3.114}$$

where

$$\phi = \tan^{-1}\left(\frac{\lambda z}{\pi w_0^2}\right) \tag{3.115}$$

and

$$r^2 = x^2 + y^2 \tag{3.116}$$

ϕ represents the phase shift between this beam and the idealized plane wave. At any point along the direction of propagation the beam profile in the x, y plane in a Gaussian with a half width (1/e point) w. w increases in size from the beam waist, w_0 at $z = 0$, according to the equation

$$w^2 = w_0^2(1 + (\lambda z/\pi w_0^2)^2) \tag{3.117}$$

$R(z)$ represents the radius of curvature of the wavefronts (or points of constant phase) and is given by the relation

$$R = z(1 + (\pi w_0^2/\lambda z)^2) \tag{3.118}$$

Both at the beam waist ($z = 0$), and at very large distances, the wavefront is flat, as in a plane wave. The contour of the beam represented by w is a hyperbola with asymptotes inclined to the axis at an angle

$$\theta = \frac{\lambda}{\pi w_0} \tag{3.119}$$

Figure 3.16 illustrates the basic parameters. w and R are real numbers belonging to a single complex variable which is given by the equation

$$\frac{1}{q(z)} = \frac{1}{R(z)} - \frac{i\lambda}{\pi w(z)^2} \tag{3.120}$$

q describes the Gaussian beam at all points along its path. If q_1 at z_1 is known then q_2 at z_2 is

$$q_2 = q_1 + (z_2 - z_1) \tag{3.121}$$

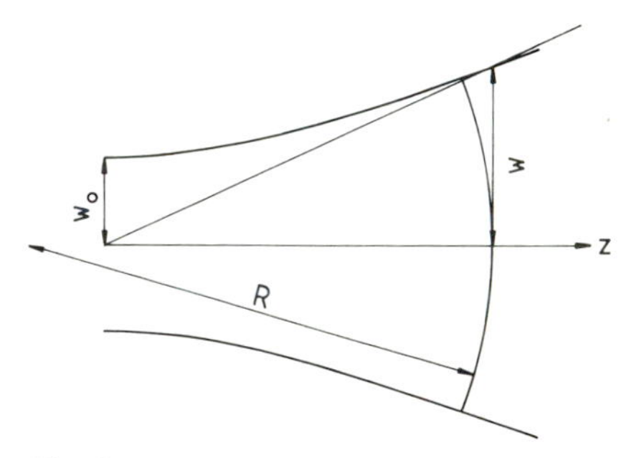

Figure 3.16. The Gaussian beam as represented by the parameter w, which is the half width of the beam measured to the point at which the intensity has fallen to 1/e of its value along the axis. R is the radius of curvature of the wavefronts and w_0 is the beam waist size.

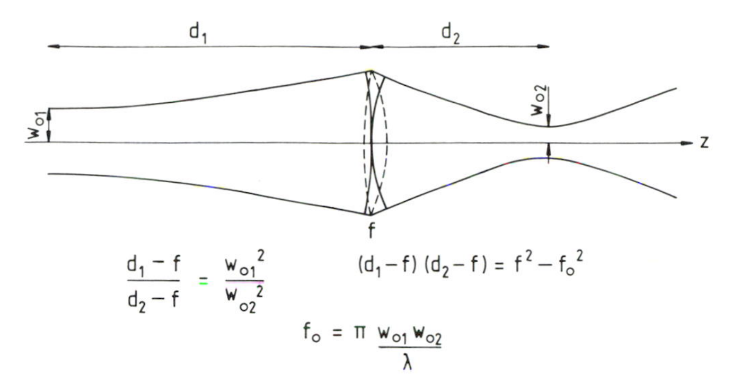

Figure 3.17. Waist to waist transport of a Gaussian beam through a thin lens, calculated for $\pi/\lambda = 1$.

This equation can be derived by a more detailed analysis of the solution to the differential equation (3.113). The beam shown in Figure 3.16 will continue to expand (in both directions!) indefinitely unless it meets some optical elements. For example, a lens would cause an alteration in the curvature of the wavefronts R as shown in Figure 3.17. If this is a convex lens of sufficient strength a new real waist can be formed at the other side of the lens. Alternatively, for a weaker or concave lens the beam after the lens will have a virtual waist on the same side of the lens. In either case the change in R at the lens is given by the Newtonian formula

$$\frac{1}{R_2} = \frac{1}{R_1} - \frac{1}{f} \tag{3.122}$$

where f is the focal length. The usual sign convention is for the radius to be positive if it is convex as viewed from $z = \infty$. Because the beam size does not alter for a thin lens the change in q is therefore

$$\frac{1}{q_2} = \frac{1}{q_1} - \frac{1}{f} \tag{3.123}$$

We can now relate the q parameter at the second waist to that at the first using Equation (3.123) and (3.121). The full formula is

$$q_{02} = \frac{(1-d_2/f)q_{01} + (d_1 + d_2 - d_1 d_2/f)}{-(q_{01}/f) + (1 - d_1/f)} \tag{3.124}$$

where d_1 and d_2 are the distances from the lens to the waists as shown in the figure. The virtue of the complex formulation can now be seen by calculating the image distances and sizes at the two waists. At these points q_1 and q_2 are the imaginary quantities

$$q_{02} = \frac{i\pi w_{02}^2}{\lambda} \tag{3.125}$$

and

$$q_{01} = \frac{i\pi w_{01}^2}{\lambda} \tag{3.126}$$

These can be substituted into (3.124) and equating the real and imaginary parts, this gives

$$\frac{d_1 - f}{d_2 - f} = \frac{w_{01}^2}{w_{02}^2} \tag{3.127}$$

and

$$(d_1 - f)(d_2 - f) = f^2 - f_0^2 \tag{3.128}$$

where

$$f_0 = \frac{\pi w_{01} w_{02}}{\lambda} \tag{3.129}$$

is usually called the characteristic length. Equation (3.128) is the same as the Newtonian formulae for point to point imaging when f_0 is zero. Physically, this condition cannot be achieved in a real beam since the occupied phase space (or emittance) must be finite.

3.11. Optical properties of laser cavities

Consider, in the first instance, how to calculate the general stability of a passive laser cavity. As far as the optics is concerned the cavity can be replaced by a system of lenses as shown in Figure 3.18. For a standing wave resonator there are two lens strengths in the infinite series corresponding to the focal lengths of the two mirrors. A complete round trip of the resonator corresponds to the transmission through a pair of these lenses. The imaging properties of any one modular element (or pair of lenses) of the infinite series can be represented by the general matrix equation

$$\begin{bmatrix} r_2 \\ r'_2 \end{bmatrix} = \begin{bmatrix} A & B \\ C & D \end{bmatrix} \begin{bmatrix} r_1 \\ r'_1 \end{bmatrix} \tag{3.130}$$

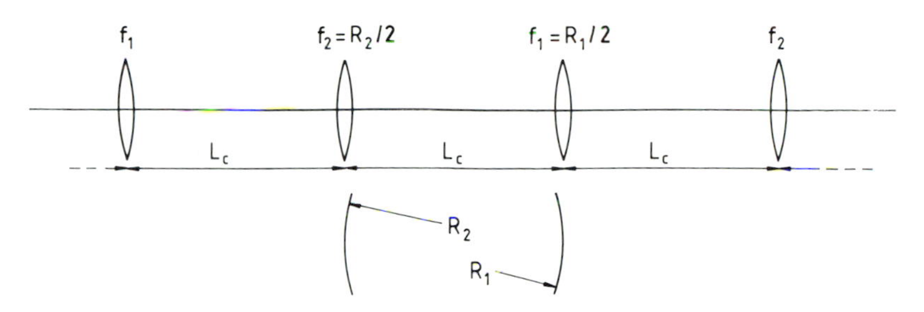

Figure 3.18. A laser cavity with two concave mirrors and its equivalent sequence of lenses.

which has the following meaning. Any paraxial ray (a ray where $\sin\theta \simeq \theta$) at a distance r_1 from the axis and with angular divergence r_1' after passing through the optical element (module), will be found at a distance r_2 with divergence r_2'. The transfer matrix *ABCD* has a determinant of unity and can be found by an elementary ray analysis. If we think of the ray travelling first a distance L_c (Figure 3.18) before passing through the lens f_1, then travelling a further distance L_c before passing through lens f_2, the ray tracing matrix (RTM) for a complete module is

$$\begin{bmatrix} A & B \\ C & D \end{bmatrix} \equiv \begin{bmatrix} 1-\dfrac{L_c}{f_1} & 2L_c-\dfrac{L_c^{\,2}}{f_1} \\ -\dfrac{1}{f_1}-\dfrac{1}{f_2}+\dfrac{L_c}{f_1 f_2} & 1-\dfrac{L_c}{f_1}-\dfrac{2L_c}{f_2}+\dfrac{L_c^{\,2}}{f_1 f_2} \end{bmatrix} \tag{3.131}$$

The repeated passage in the laser cavity is equivalent to a large number ($n \to \infty$) of successive operations of the transfer matrix for a complete round trip. For n operations this can be simplified using Sylvester's theorem to

$$\begin{bmatrix} A & B \\ C & D \end{bmatrix}^n = \frac{1}{\sin\theta}\begin{bmatrix} A\sin n\theta-\sin(n-1)\theta & B\sin n\theta \\ C\sin n\theta & D\sin n\theta-\sin(n-1)\theta \end{bmatrix} \tag{3.132}$$

where

$$\cos\theta = \tfrac{1}{2}(A+D) \tag{3.133}$$

The stability criteria can now be easily seen in this last equation. If the RHS obeys the following inequality

$$-1 < \tfrac{1}{2}(A+D) < 1 \tag{3.134}$$

then θ is a real quantity and the cavity is stable. The beam profile along the infinite series repeats itself after a length equivalent to $2L_c$, where L_c is the cavity separation. On the other hand, if the inequality is not obeyed the functions are hyperbolic and the cavity is unstable. The matrix then changes indefinitely with increasing n and the beam expands until it spills over the edges of the mirrors. In an unstable resonator any ray which has a finite displacement from the axis (r) or angular divergence (r') is lost from the cavity. The ray $r = 0, r' = 0$, which is the symmetry axis, is of course an exception. For a standing wave laser we can substitute the values of A and D from Equation (3.131) into the inequality (3.134) to give

$$0 < (1 - L_c/R_1)(1 - L_c/R_2) > 1 \tag{3.135}$$

for the stability criterion, where

$$f_1 = \frac{R_1}{2} \quad \text{and} \quad f_2 = \frac{R_2}{2} \tag{3.136}$$

This is plotted in Figure 3.19. It can be seen that both the plane parallel and the confocal resonator are on the borderline of stability. The former geometry is now seldom used in lasers, whilst the latter can be brought into the stability region by a small change in the separation L_c.

We can now look at the propagation of a Gaussian beam in a stable passive cavity. This can be done in a very general way using the equivalent sequence of lenses as before. Any type of laser is applicable to this method including ring

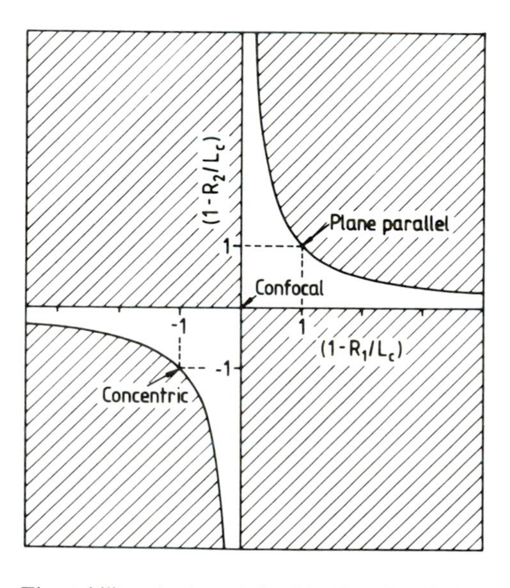

Figure 3.19. The stability criteria as defined by Equation (3.134). The shaded areas correspond to unstable resonators.

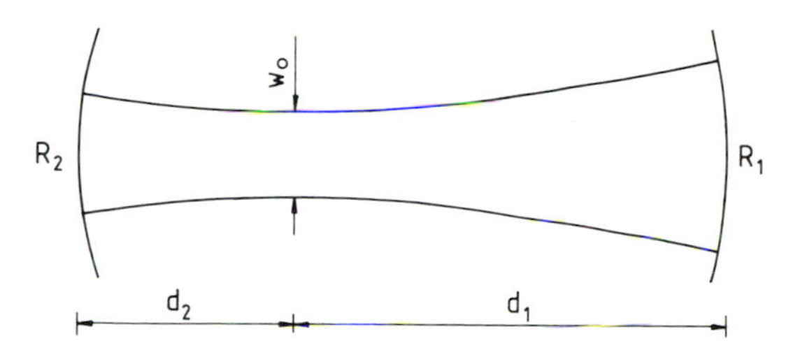

Figure 3.20. A Gaussian beam in a laser cavity with concave mirrors for $\pi/\lambda = 1$.

lasers and those with intracavity optics. However, a simpler analysis, for a standing wave laser, is to adjust the parameters of a single Gaussian beam so that the curvature of the wavefronts is matched at both mirrors. Figure 3.20 shows this for a laser with two concave mirrors where a waist is formed at some point between the mirrors. When one mirror is convex the waist must be external to the cavity. To calculate the size and position of the waist the Equations (3.128) and (3.129) are used. Since the 'image' and 'object' distances are equal for both mirrors then

$$(d_2 - f_2)^2 = f_2^2 - (\pi w_0^2/\lambda)^2 \tag{3.137}$$

$$(d_1 - f_1)^2 = f_1^2 - (\pi w_0^2/\lambda)^2 \tag{3.138}$$

and

$$d_1 + d_2 = L_c \tag{3.139}$$

A solution for the confocal geometry is now simple, since $d_1 = d_2 = L_c/2$, and therefore

$$w_0^2 = \frac{f_2 \lambda}{\pi} = \frac{L_2 \lambda}{2\pi} \tag{3.140}$$

The beam spot on the mirrors which is found from Equation (3.117) is

$$w_1^2 = w_2^2 = \frac{L_c \lambda}{\pi} \tag{3.141}$$

General solutions to the three equations can be found for d_1, d_2 and the beam waist, w_0. The latter defines the Gaussian beam both inside and outside the cavity. Subtracting (3.138) from (3.137) and using (3.139) gives two equations for d_2 and d_1 which have solutions

$$d_2 = \frac{L_c(R_1 - L_c)}{R_1 + R_2 - 2L_c} \tag{3.142}$$

and

$$d_1 = \frac{L_c(R_2 - L_c)}{R_1 + R_2 - 2L_c} \tag{3.143}$$

These can be substituted into (3.137) or (3.138) to give

$$w_0^4 = \left(\frac{\lambda}{\pi}\right)^2 \frac{L_c(R_1 - L_c)(R_2 - L_c)(R_1 + R_2 - L_c)}{(R_1 + R_2 - 2L_c)^2} \tag{3.144}$$

for the beam waist size. The beam spots on the mirrors are then given by the equations

$$w_1^4 = \left(\frac{\lambda R_1}{\pi}\right)^2 \frac{L_c(R_2 - L_c)}{(R_1 - L_c)(R_1 + R_2 - L_c)} \tag{3.145}$$

and

$$w_2^4 = \left(\frac{\lambda R_2}{\pi}\right)^2 \frac{L_2(R_1 - L_c)}{(R_1 - L_c)(R_1 + R_2 - L_c)} \tag{3.146}$$

w_0, w_1 and w_2 are critical parameters in any laser design. The radial dimension of the cylinder containing the amplifying medium (e.g. the discharge tube in a gas laser) must be larger than w_0 but not so large that pumping power is wasted, whilst the mirrors must have diameters significantly in excess of w_1 and w_2.

3.12. *Mode frequencies*

In deriving Equation (3.92) for the allowed frequencies the transverse mode structure was ignored. The total phase change is made up of a contribution† from transverse and longitudinal components so the mode frequencies depend also on the complex part of the attenuation coefficient. If this is written

$$\gamma = |\gamma| \exp(i\beta) \tag{3.147}$$

† The full solution for the electric field can be separated out into a product of an x, y dependent part and a z dependent part, so the phases are just added.

then the phase change from the x, y-dependent (or R-dependent) part is β, and the allowed mode frequencies (angular) are

$$\omega_{q\beta} = \frac{c(q\pi + \beta)}{L_c} \tag{3.148}$$

An analytical expression is possible for β when the spot radius is much smaller than the aperture of laser mirrors. The integral (3.107) can then be used to calculate γ explicitly. γ depends on the type of the transverse mode as well as the details of the geometry. This geometry effect can be formulated in terms of the radius of curvature of each mirror and the cavity length using the parameters

$$g_1 = 1 - \frac{L_c}{R_1} \tag{3.149}$$

and

$$g_2 = 1 - \frac{L_c}{R_2} \tag{3.150}$$

where R_1 and R_2 are the mirror radii. The mode frequencies for circular mirrors are then given by the equation

$$\omega_{plq} = c(q\pi + (2p + l + 1)\cos^{-1}(g_1 g_2)^{1/2})/L_c \tag{3.151}$$

When the mirrors are rectangular, $2p$ and l are replaced by the integers m and n. For a plane-parallel cavity, $g_1 = g_2 = 1$, and the frequencies are the same as our simple analysis (Equation 3.92). The confocal geometry $g_1 = g_2 = 0$ gives

$$\omega_{plq} = c\pi(2q + (2p + l + 1))2L_c \tag{3.152}$$

Here there is considerable degeneracy but the mode spacing† in all cases is

$$\Delta\omega = \frac{\pi c}{L_c} \tag{3.153}$$

In almost all practical cases the interest is primarily for $(\mathrm{TEM})_{00}$ modes where Equations (3.151) and (3.152) differ by a factor of 1/2. It should be borne in mind that we are still discussing passive cavities and the effects of anomalous dispersion have not been considered. The mode spacings will be altered by small amounts (mode pulling) dependent on their position relative to the small signal gain profile.

† Assuming $\eta = 1$. For a laser shown in Figure 3.8, $L_c = \sum_i \eta_i \bar{L}_i$ where L_i are the lengths of the regions of different refractive index.

3.13. Saturation effects and laser output power

We now come to the point at which we can collect together the previous sections and discuss an entire laser oscillator. Unfortunately, the rigorous quantum mechanical theories, or the semiclassical theory of Lamb† are beyond the scope of this book. Here we shall be content to consider the physics by an extension of the techniques used earlier for the analysis of a travelling wave amplifier. This treatment is associated with Bennett‡. It uses intensity rather than amplitudes and so neglects any phenomena associated with the phase of the wave. Topics such as spatial hole burning and mode pulling can, thus, only be discussed in a qualitative manner.

In the first instance, consider how the laser oscillator is different from the travelling wave amplifier. Two aspects are most important. Firstly, the system may oscillate at several frequencies determined by the passive response of the cavity. The number of frequencies will depend on the mode spacing and the total width of the gain profile. Secondly, the standing wave in the cavity is composed of two components travelling in opposite directions, each of which will contribute to the saturation characteristics. When the transition is dominated by Doppler broadening, these two waves interact mainly with different groups of atoms (Figure 3.4) except when the frequency is near to the line centre. At this point the output power decreases because the gain saturates more rapidly. This effect is known as the Lamb dip, and a simple theory of this is given later.

Another difficulty arises because of the intensity variation normal to the direction of propagation which will produce an uneven variation in the saturation of the amplifying medium. We will choose to neglect this and assume that the output power can be approximately obtained by multiplying the average intensity, from the following calculations, by the area of the beam inside the Gaussian half height points.

The easiest case is for a homogeneously broadened system. If the round trip loss is small then the intensity can be found from (3.70) but with the gain, α, duly readjusted to account for beam propagating in both directions. Denoting I_+ and I_- for the beam propagating in the two directions then Equations (3.70) and (3.37) give (ignoring spatial hole burning!)

$$\alpha = \frac{\alpha_0}{1+\bar{f}(I_+ + I_-)/I_s} = \frac{\delta_c}{2L} \tag{3.154}$$

where α_0 is the small signal gain (3.29). Here it is assumed that only a single frequency propagates. This is always true for a homogeneously broadened transition because the gain profile is reduced by saturation until only a single

† W.E. Lamb, *Phys. Rev.* **A134** (1964) 1429.
‡ W.R. Bennett, *J. Appl. Opt. Suppl., Optical Masers* (1962).

cavity mode has sufficient gain to overcome the losses. The total cavity loss in a round trip, δ_c, will have three components representing the losses due to scattering, diffraction and output. This can be written

$$\delta_c = 2L\beta + (1 - 2\gamma\gamma^*) + (1 - R_1 R_2) \tag{3.155}$$

The middle term which is the diffraction loss for two passes of the cavity, applies only where $\gamma \ll 1$. β is the fractional loss in intensity per unit length due to scattering and again this assumes $\beta \ll 1$. Other terms such as scattering from imperfect Brewster windows can be simply added to this equation. Noting that when R_1 and R_2 are close to unity the intensities I_+ and I_- are nearly equal for all values of z, then

$$I_+(L) \simeq \frac{I_s}{2\bar{f}}\left(\frac{2L\alpha_0}{\delta_c} - 1\right) \tag{3.156}$$

and the output intensity from both mirrors is

$$I_{\text{out}} \simeq I_+(1 - R_1 R_2) \tag{3.157}$$

so the output power is approximately this intensity multiplied by the beam 'area'. Usually the cavity mode is sufficiently close to line centre so that $\bar{f} = 1$ and

$$\alpha_0(\omega) \simeq \alpha_0(\omega_{12}) = B\hbar\omega(n_2^0 - (g_2/g_1)n_1^0)/\pi\gamma_t c \tag{3.158}$$

where γ_t is the half width of the homogeneous profile. Combining (3.157), (3.155) and (3.156) gives

$$I_{\text{out}} = \frac{TI_s}{2}\left(\frac{g_0}{S + D + T} - 1\right) \tag{3.159}$$

where

$$S = 2L\beta \tag{3.160}$$

is the round trip scattering loss and

$$D = 1 - 2\gamma\gamma^* \tag{3.161}$$

is the round trip diffraction loss and

$$T = 1 - R_1 R_2 \tag{3.162}$$

is the total percentage of power transmitted through the mirrors. g_0 is the total round-trip small signal gain, and is given by the relation

$$g_0 = 2L\alpha_0 \tag{3.163}$$

Normally the laser is designed so that power is only extracted from one mirror ($R_1 = 1$). Equation (3.15) can be differentiated with respect to the output transmission (or coupling) T to give

$$\frac{\partial I_{\text{out}}}{\partial T} = \frac{I_s}{2}\left(\frac{g_0}{A+T} - 1\right) - \frac{I_s T g_0}{2(A+T)^2} \tag{3.164}$$

where

$$A = D + S \tag{3.165}$$

represents the total 'internal' losses. From Equation (3.164) the output power is a maximum when the output coupling is

$$T = (g_0 A)^{1/2} - A \tag{3.166}$$

The output intensity is plotted in Figure 3.21 for various assumed losses.

Equation (3.159) was derived by assuming that the gain, g_0, and losses, δ_c per round trip were small. For a more exact analysis the spatial variation of intensity along the amplifying medium needs to be calculated. Consider the

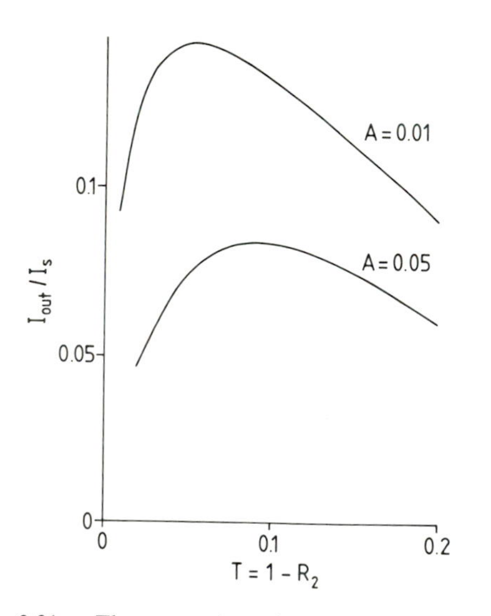

Figure 3.21. The output intensity for $g_0 = 0{\cdot}4$ as a function of the output coupling. Curves are plotted for round trip internal losses of 1% and 5%.

approximation (in a well designed laser!) that the diffraction losses are small enough to be considered as taking place just at the mirrors. The two equations determining the variation in intensity can be obtained by combining (3.154) and (3.28). These are, for $\bar{f} = 1$,

$$\frac{\partial I_+}{\partial z} = \frac{\alpha_0 I_+}{1 + (I_+ + I_-)/I_s} - \beta I_+ \tag{3.167}$$

and

$$-\frac{\partial I_-}{\partial z} = \frac{\alpha_0 I_-}{1 + (I_+ + I_-)/I_s} - \beta I_- \tag{3.168}$$

The minus sign in front of $(\partial I_- / \partial z)$ indicates that the reverse wave increases with decreasing z. Firstly, it can be seen that

$$\frac{1}{I_+}\left(\frac{\partial I_+}{\partial z}\right) = -\frac{1}{I_-}\left(\frac{\partial I_-}{\partial z}\right) \tag{3.169}$$

or

$$\frac{\partial}{\partial z}(\log(I_+ I_-)) = 0 \tag{3.170}$$

giving

$$I_+ I_- = k^2 \tag{3.171}$$

where k^2 is a constant independent of z. We now solve the two coupled equations with appropriate boundary conditions. This can only be done analytically for the approximation that all the internal losses including β are lumped together at the ends. We expect therefore that the approximation is good when the output transmission losses are much larger than the distributed losses. On the other hand when the output coupling is small, and comparable with the distributed losses, the linear variation implicit in (3.69) can be used. It is therefore only necessary to consider the more complicated equations, (3.167) and (3.168), when the round trip gain is very high so that the optimum coupling is also large.

The boundary conditions at the mirrors, which are the same as those at the ends of the amplifying medium, can then be written as

$$I_-(L) = R_2' I_+(L) \tag{3.172}$$

and

$$I_+(0) = R_1' I_-(0) \tag{3.173}$$

where R_2' and R_1' are the reflection coefficients modified to account for the small extra distributed losses. Dividing these losses equally between the two mirrors gives

$$R_2' = R_2 - A/2 \tag{3.174}$$

and

$$R_1' = R_1 - A/2 \tag{3.175}$$

A must be considerably smaller than the total output coupling since the second term due to scattering (or diffraction had this been included) has to be neglected. Equation (3.171) can be combined with (3.172) and (3.173) to give the following relations

$$I_+(L)^2 = k^2/R_2' \tag{3.176}$$

$$I_+(0)^2 = k^2 R_1' \tag{3.177}$$

$$I_-(L)^2 = k^2 R_2' \tag{3.178}$$

and

$$I_-(0)^2 = k^2/R_1' \tag{3.179}$$

Assuming β is zero the two equations (3.167) and (3.168) are integrated directly to give

$$\ln\left[\frac{I_+(L)}{I_+(0)}\right] + \frac{(I_+(L) - I_+(0))}{I_s} + \frac{k^2(1/I_+(0) - 1/I_+(L))}{I_2} = \alpha_0 L \tag{3.180}$$

and

$$\ln\left[\frac{I_-(L)}{I_-(0)}\right] + \frac{(I_-(L) - I_-(0))}{I_s} + \frac{k^2(1/I_-(0) - 1/I_-(L))}{I_s} = -\alpha_0 L \tag{3.181}$$

Solving for $I_+(L)$ by subtracting (3.181) from (3.180), and then using the boundary conditions (3.172) and (3.173), together with the equations (3.176) to (3.179), gives

$$I_+(L) = \frac{I_s(\alpha_0 L - (1/2)\ln(1/R_2' R_1'))}{1 - R_2' + (R_2'/R_1')^{1/2}(1 - R_1')} \tag{3.182}$$

Considering that the beam is extracted only from mirror 2, by putting $R_1 = 1$, then the output intensity is

$$I_{\text{out}} = \frac{(1-R_2)I_s(\alpha_0 L - (1/2)\ln(1/R_2' R_1'))}{(1-R_2') + (R_2'/R_1')^{1/2}(1-R_1')} \tag{3.183}$$

This result is exact for zero internal losses. It is also approximately true when these losses are small, even when the output coupling is high. When both are small it approximates to expression (3.159).

When the gain profile is dominated by Doppler broadening the calculation is considerably more difficult. The laser can oscillate at several different frequencies spanning the inhomogeneous profile and the forward and reverse waves will not, in general, contribute equally to the saturation characteristics in the manner of Equation (3.154). Consider the case of a standing wave laser oscillating in a single cavity mode. (When more than one mode is present then an intra-cavity etalon can be used to suppress all but a single mode. This is in fact a desirable condition for laser applications in spectroscopy or optics where a single frequency or a very coherent beam, is required.) If the single mode is far enough away from the centre of the gain profile, and the homogeneous linewidth is relatively small, then the output intensity can be obtained by equating the travelling wave gain, (3.56), to the total distributed losses. Thus, the gain for a monochromatic beam of intensity I_ω is

$$\alpha(\omega) = \frac{\alpha_0(\omega)}{(1+I_\omega/I_s)^{1/2}} = \frac{\delta_c}{2L} \tag{3.184}$$

where

$$I_\omega = I_+ = I_- \tag{3.185}$$

Here we are acknowledging the fact that the return wave burns a different hole in the velocity distribution (Figure 3.21) when the cavity mode is away from the line centre ω_{12}, but these two holes are equidistant from the centre of the velocity distribution and have the same size. Note that α_0 now refers to the inhomogeneous case and will have a Gaussian shape (or a Voigt profile for the general case) as determined by (3.56). The cavity intensity is therefore

$$I_\omega = I_s\left[\left(\frac{2\alpha_0 L}{\delta_c}\right)^2 - 1\right] \tag{3.186}$$

and the output at mirror 2 is therefore

$$I_{\text{out}} = (1-R_2)I_s\left[\left(\frac{2\alpha L}{\delta_c}\right)^2 - 1\right] \tag{3.187}$$

Equation (3.183) is approximately true when the cavity losses are small. For a high gain system, with a matched output coupling, an analysis similar to that for the homogeneous case is required. This is left as an exercise for the reader.

Now consider the more general case, when both waves are included, but, again, only one frequency is present. The total population inversion as a function of the velocity of the atoms is obtained from Equation (3.47), but including a term for both the forward and reverse wave. In the case of the reverse wave the sign of the velocity is changed so that

$$n_2(v)-(g_2/g_1)n_1(v) = \frac{(n_2^0-(g_2/g_1)n_1^0)n_1^0)\rho(v)}{1+(I_\omega/I_s)(\bar{f}_{2+}(\omega,v)+\bar{f}_{2-}(\omega,v))} \tag{3.188}$$

where

$$\bar{f}_{2+} = \frac{\gamma_t^2}{(\omega_{12}-\omega(1-v/c))^2+\gamma_t^2} \tag{3.189}$$

and

$$\bar{f}_{2-} = \frac{\gamma_t^2}{(\omega_{12}-\omega(1+v/c))^2+\gamma_t^2} \tag{3.190}$$

Implicit in Equation (3.188) is that I_ω is the same in the return wave as the forward wave for all values of z. This is approximately true for a low loss system. Here the power in the cavity is much larger than the output power and the change in intensity for a round trip is a small fraction of the total cavity intensity. Notice that $\rho(v)$ is the same for both the forward and reverse waves since the Gaussian is only dependent on v^2. To obtain the gain we now have to multiply the inversion density by the cross section $\sigma(\omega, v)$ in the manner of Equation (3.50). However the cross section will be different for the forward and reverse waves because of the change in $\bar{f}_2(\omega, v)$. The average gain on a round trip is therefore

$$\bar{\alpha}(\omega, v) = (n_2^0-(g_2/g_1)n_1^0)\sigma(\omega_{12})\rho(v)\frac{(\bar{f}_{2+}(\omega,v)+\bar{f}_{2-}(\omega,v))/2}{1+(I_\omega/I_s)(\bar{f}_{2+}(\omega,v)+\bar{f}_{2-}(\omega,v))} \tag{3.191}$$

where $\sigma(\omega_{12})$ is given by (3.26). The total gain for all velocities of atoms is then

$$\bar{\alpha}(\omega) = (n_2^0-(g_2/g_1)n_1^0)\sigma(\omega_{12})\int_{-\infty}^{+\infty}\frac{\rho(v)(\bar{f}_{2+}(\omega,v)+\bar{f}_{2-}(\omega,v))/2}{1+(I_\omega/I_s)(\bar{f}_{2+}(\omega,v)+\bar{f}_{2-}(\omega,v)))}\,dv \tag{3.192}$$

As before, this cannot be solved analytically unless the homogeneous width is much narrower than the Doppler broadening. When this is true, ρ can be replaced by its average value at the centre of the Lorenzian and removed from the integral. To proceed further we made the assumption that I_ω is much less than I_s, so that the bottom line can be replaced by a binomial expansion up to the lowest order in I_ω/I_s. This is only a good approximation when the round trip (small signal) gain g_0 is not much bigger than the total losses. Equation (3.192) then becomes

$$\bar{\alpha}(\omega) = (n_2^0 - (g_2/g_1)n_1^0)\sigma(\omega_{12})\rho(v_r)$$
$$\times\left[\frac{1}{2}\int_{-\infty}^{+\infty}(\bar{f}_{2+} + \bar{f}_{2-})\,dv - \frac{I_\omega}{2I_s}\int_{-\infty}^{+\infty}(\bar{f}_{2+}^2 + \bar{f}_{2-}^2)\,dv - \frac{I_\omega}{2I_s}\int_{-\infty}^{+\infty}2\bar{f}_{2+}\,\bar{f}_{2-}\,dv\right] \tag{3.193}$$

where $\rho(v_r)$ is the normalized Gaussian (2.178) at $v_r = c(1-\omega_{12}/\omega)$. The integrals in (3.193) are a little frightening but can be solved using (3.53) together with the useful convolution

$$\int_{-\infty}^{+\infty}\frac{1}{(1+x^2)}\frac{1}{(1+(x-a)^2)}\,dx = \frac{2\pi}{a^2+4} \tag{3.194}$$

Now the first two integrals in the above expansion correspond to our previous result for the travelling wave (3.52), when the denominator is expanded by the binomial theorem. The last integral corresponds to the 'interference' term which is the case of the Lamb dip. Performing the integration, and using the results (3.55) and (3.57), one obtains

$$\bar{\alpha}(\omega) = \alpha_0(\omega)\left[1 + \frac{I_\omega}{I_s} - \left(1 + \frac{\gamma_t^2}{(\omega_{12}-\omega)^2 + \gamma_t^2}\right)\right]^{-1} \tag{3.195}$$

where $\alpha_0(\omega)$ is the small signal gain, which is given by Equation (3.56) when $I_\omega = 0$. Thus

$$\alpha_0(\omega) = (n_2^0 - (g_2/g_1)n_1^0)\frac{B\hbar\omega g(\omega)}{\eta c} \tag{3.196}$$

and $g(\omega)$ is the Gaussian (3.57). It is now a simple matter to obtain the cavity intensity by equating the cavity losses to the gain. This gives

$$I_\omega = 2I_s\left(\frac{2\alpha_0 L}{\delta_c} - 1\right)\left(1 + \frac{\gamma_t^2}{(\omega_{12}-\omega)^2 + \gamma_t^2}\right)^{-1} \tag{3.197}$$

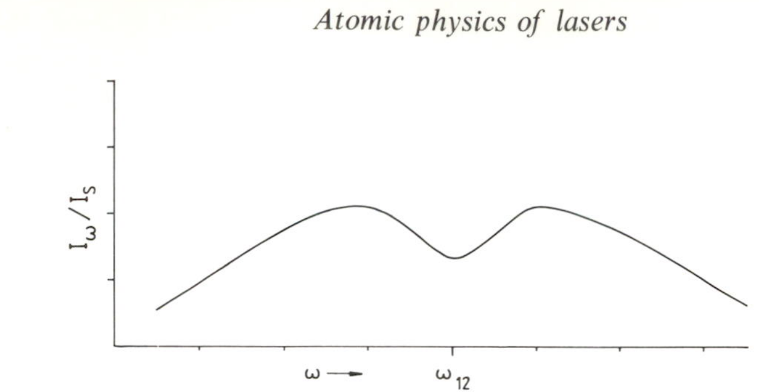

Figure 3.22. The output intensity as a function of frequency showing the Lamb dip.

and the output intensity is just this function multiplied by the output coupling.

Figure 3.22 plots the cavity intensity (3.197) as a function of frequency. Close to the line centre a dip is observed which will have a Lorenzian profile when the homogeneous linewidth is much smaller than the Doppler broadening. If the frequency is more than a few homogeneous linewidths away from line centre then the equation approximates to (3.186) when $I_\omega/I_s \ll 1$.

Note carefully that this analysis is fairly restrictive on the conditions which need to apply before an analytical result can be obtained. In other circumstances, say when I_ω/I_s is large, or when γ_t is a significant fraction of γ_I, one must resort to numerical techniques. These produce gain profiles which have essentially the same features as our simplified analysis. The theory outlined here is, of course, equally applicable to saturated absorption spectroscopy.

Bibliography

Allen, L., 1969, *Essentials of Lasers* (Oxford: Pergamon Press).
Birnbaum, G., 1964, *Optical Masers* (New York: Academic Press).
Corney, A., 1977, *Atomic and Laser Spectroscopy* (Oxford University Press).
Demtroder, W., 1981, *Laser Spectroscopy* (Berlin: Springer-Verlag).
Hecht, E. and Zajac, A., 1974, *Optics* (Addison-Wesley).
Loudon, R., 1983, *The Quantum Theory of Light* (2nd edition) (Oxford University Press).
Maitland, A. and Dunn, M.H., 1969, *Laser Physics* (Amsterdam: North Holland Publishing Company).
Verdeyen, J.T., 1981, *Laser Electronics* (New Jersey: Prentice-Hall).

CHAPTER 4

special topics

4.0. Introduction

This chapter is essentially a continuation of the previous one and expands the basic ideas of light amplification, considering in particular three and four level systems. Topics such as mode locking and Q-switching are also discussed. As before the starting point is the gain coefficient which is represented by Equation (3.10) with the sign altered. The small signal gain can always be written

$$\alpha^0(\omega) = (n_2^0 - (g_2/g_1)n_1^0)\frac{B\hbar\omega\, f(\omega)}{\eta c} \tag{4.1}$$

where $f(\omega)$ will have a Voigt profile for the completely general case, but, in many situations, it can be approximated by either a normalized Lorenzian (3.4) or Gaussian (3.57). Note carefully that this equation is quite general and applies independently of any external influences such as pumping rates. Now previously we considered in detail the saturation characteristics for both homogeneous (Section 3.2), and Doppler (Section 3.3) broadened transitions, but in doing so we assumed that the upper level could be populated at a rate dependent on the pumping power and not on the ground state population. Now we consider what happens if the complete cycle of transitions is involved.

4.1. Three level laser system

Figure 4.1 shows a simplified schematic diagram for a three level system, which is typical of some doped insulator lasers (e.g. ruby). In these types of lasers optical pumping is used to populate a broad energy level E_3, which then undergoes radiationless decay to E_2 with a lifetime τ_{32}. The electromagnetic decay from the level E_2 back to the ground state then provides the lasing transition, and we must have n_2 greater than n_1 for positive amplification. Now certain things are fairly obvious without any algebra. Firstly, it is essential that the relaxation time τ_{32} is much shorter than the spontaneous decay of the level

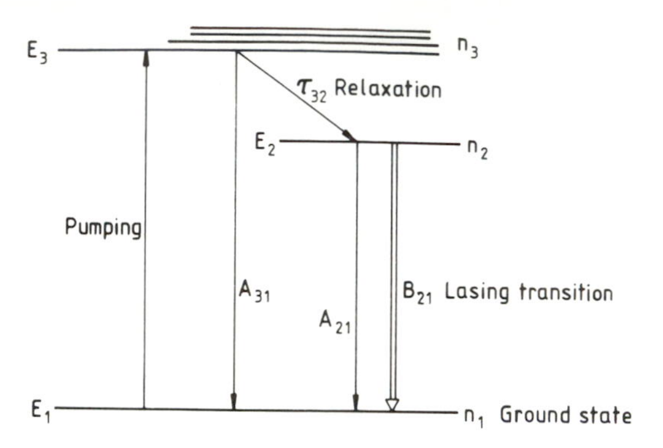

Figure 4.1. Diagram of a three level laser system.

E_2. Even if the pumping were to completely saturate the level E_3, its population density $(g_1/g_3 n_3)$ could not exceed the ground state population. To produce a population inversion between level 2 and 1 demands that atoms are effectively 'stored' in level 2 and this is brought about by ensuring that

$$\tau_{32} \ll \tau_{21} \tag{4.2}$$

or in terms of the Einstein coefficients

$$\tau_{32} \ll 1/A_{21} \tag{4.3}$$

Of course, when the laser is operating the effective lifetime of level 2 will be considerably reduced by saturation. Secondly, it is important for the overall efficiency if the spontaneous decay from level 3 to the ground state is much less probable than the radiationless transition. In this way all the pumping goes into producing the inversion.

A qualitative analysis is similar to Section 3.2 except that there is an extra conservation equation

$$n_0 = n_1 + n_2 + n_3 \tag{4.4}$$

where n_0 is the total number density of atoms. The rate equations, for a homogeneously broadened transition and single mode propagation, are

$$\frac{dn_1}{dt} = -R + n_3 A_{31} + n_2 A_{21} + B_{21} Wf\left(n_2 - \frac{g_2}{g_1}\right) n_1 \tag{4.5}$$

$$\frac{dn_2}{dt} = n_3 S_{32} - n_2 A_{21} - B_{21} Wf\left(n_2 - \frac{g_2}{g_1}\right) n_1 \tag{4.6}$$

and

$$\frac{dn_3}{dt} = R - n_3 A_{31} - n_3 S_{32} \tag{4.7}$$

where f is the broadened lineshape, centred at ω_{12}, and the refractive index has been incorporated into the B coefficient. Now, if the system is dominated by inhomogeneous effects, as, for example, in a low-temperature ruby laser these equations cannot be used by themselves. However, for simplicity, we avoid repeating the more complicated analysis of Section 3.3. In contrast to Section 3.2 the equations have been written in terms of the A coefficients which are related to the lifetimes (neglecting any non-radiative transfer) by the equations

$$\tau_{21} = \frac{1}{A_{21}} \tag{4.8}$$

and

$$\tau_{31} = \frac{1}{A_{31}} \tag{4.9}$$

The decay rate S_{32} is related to the lifetime τ_{32} by the relation

$$\tau_{32} = \frac{1}{S_{32}} \tag{4.10}$$

W is the energy density of the amplified beam† which can be expressed in terms of the intensity using Equation (3.1), and R is the pumping rate which can be expressed in terms of the Einstein coefficient B_{13} and the energy density of the pumping source. If this source is slowly varying in comparison to the absorption linewidth then

$$R = Wp(\omega_{13}) B_{13}(n_1 - (g_1/g_3) n_3) \tag{4.11}$$

where $Wp(\omega_{13})$ is the energy density at ω_{13}. Quite often, however, the pumping consists of a flash tube whose spectral intensity is a series of narrow lines. In the event that these lines are narrower than level 3, the pumping rate is given by the integral

$$R = (n_1 - (g_1/g_3) n_3) B_{13} \int_0^\infty h(\omega) Wp(\omega)\, d\omega \tag{4.12}$$

† Strictly speaking we should include the extra term from the pumping source in this energy density.

where $h(\omega)$ is the normalized absorption lineshape and $Wp(\omega)$ is the measured spectral output of the pump. Alternatively, for black body radiation the formula (2.15) can be used.

Equations (4.4) to (4.7) together with (4.11) can now be used to determine the number densities in the levels (and hence the gain coefficient) as a function of time, given that R, which may also be time dependent, is known and the number densities at time zero are also known. If the level spacings are much larger than the thermal excitations in the crystal (kT) then the boundary condition at zero time is for all the atoms to be in the ground state. A quantitative study of the time dependent behaviour will not be considered here, although we will consider the problem in a qualitative manner when discussing Q switching in Section 4.4. The steady state solutions are when all the derivatives on the LHS are equal to zero and the pumping rate (4.11) is constant. From the four equations (4.4) to (4.7), together with (4.11), the number densities in the three levels can be obtained as a function of n_0, Wp, W, the A and B coefficients, and the decay rate S_{32}. The values of n_2 and n_1 can then be used in Equation (3.10) to determine the gain coefficient. These solutions are very lengthy and only readers with algebraic stamina are recommended to follow them through! Here the approximations (4.2) and (4.3), which apply for example in the case of a ruby laser, will be used to simplify the solutions. Using Equation (4.11), the two steady state equations corresponding to (4.7) and (4.6), are

$$WpB_{13}n_1 = (A_{31} + S_{32} + WpB_{13}(g_1/g_3))n_3 \tag{4.13}$$

and

$$WfB_{21}(g_2/g_1)n_1 = (WB_{21}f + A_{21})n_2 - S_{32}n_3 \tag{4.14}$$

Elimination of n_3 gives

$$(WfB_{21} + A_{21})n_2 = \left(Wf(g_2/g_1)B_{21} + \frac{WpB_{13}S_{32}}{A_{31} + S_{32} + WpB_{13}(g_1/g_3)}\right)n_1 \tag{4.15}$$

Now let us take the specific case of a ruby laser. Here $S_{32} \simeq 100A_{31}$ and even with the most intense pump source WpB_{13} is much less than S_{32}. Equation (4.15) is therefore approximately

$$n_2 = \frac{(Wf(g_2/g_1)B_{21} + WpB_{13})}{(WfB_{21} + A_{21})}n_1 \tag{4.16}$$

Also S_{32} is approximately 10^5 times larger than A_{21} so the population of level three is always small, and a good approximation is

$$n_0 = n_2 + n_1 \tag{4.17}$$

Equations (4.16) and (4.17) are then used to obtain the population inversion density. This is just

$$n_2-(g_2/g_1)n_1 = \frac{n_0(WpB_{13}-(g_2/g_1)A_{21})}{WfB_{21}((g_2/g_1)+1)+WpB_{13}+A_{21}} \tag{4.18}$$

In the case of a ruby laser $g_2 = g_1$, so we have the fairly obvious conclusion that to obtain a population inversion the initial pumping rate, $n_0\, WpB_{13}$, must exceed the decay rate $n_0\, A_{21}$. Notice that if the pumping source has a significant energy density at ω_{12} then Equation (4.18) should strictly be

$$n_2-(g_2/g_1)n_1 = \frac{n_0(Wp(\omega_{13})B_{13}-(g_2/g_1)A_{21})}{WfB_{21}((g_2/g_1)+1)+Wp(\omega_{13})B_{13}+Wp(\omega_{12})(1+(g_2/g_1))B_{21}+A_{21}} \tag{4.19}$$

but because B_{21} is much smaller than B_{31} this term can be safely ignored. From Equations (4.18) and (3.10) the gain coefficient is

$$\alpha = \frac{n_0(WpB_{13}-(g_2/g_1)A_{21})}{WfB_{21}((g_2/g_1)+1)+WpB_{13}+A_{21}}(B_{21}\, f\hbar\omega/c) \tag{4.20}$$

which can be written in the form of (3.27) as

$$\alpha = \frac{(R_0\tau_{21}-(g_2/g_1)n_0)\sigma_{21}}{1+R_0(\tau_{21}/n_0)+(I\sigma_{21}/\hbar\omega)((g_2/g_1)+1)\tau_{21}} \tag{4.21}$$

where R_0 is the initial pumping rate, defined by the equation

$$R_0 = n_0\, W_p B_{13} \tag{4.22}$$

The gain coefficient including inhomogeneous effects can be calculated by integrating the gain profile (4.20) over all the classes of atoms.

We can see that the form of Equation (4.21) differs from (3.27) in respect of the second term in the denominator. If the losses are ignored then $R_0\tau_{21}$ has to exceed n_0 for lasing action. The middle term reduces the gain even without the output saturation, which is represented by the last term in the denominator. In fact, the three level system is fundamentally inefficient when compared to a four level one (which in many cases can be analysed in terms of Equation (3.27)). The main problem is that the pumping rate has to be very high because the lower level of the lasing transition is the ground state, in which all the atoms initially reside, whereas in other schemes, the lower level is largely unoccupied until pumping begins. A typical calculation of the intensities necessary to give sustained lasing action in a ruby crystal will illustrate this. Let us assume that the system is being pumped by a continuous black body spectrum which might be achieved in practice by surrounding the crystal with a large number of tungsten filament bulbs. The energy density W is then given by the Planck formula (2.15) and the condition for lasing action, ignoring any losses, is

$$\frac{8\pi \nu_{31}{}^{2} \hbar \nu_{31}}{c^{3}(\exp(h\nu_{31}/kT)-1)} > \frac{A_{21}}{B_{13}} \tag{4.23}$$

which can be written using (2.24) and (2.25) as

$$\frac{1}{\exp(h\nu_{31}/kT)-1)} > \frac{g_1 A_{21}}{g_3 A_{31}} \equiv \frac{g_1 \tau_{31}}{g_3 \tau_{21}} \tag{4.24}$$

The critical temperature in the absence of any losses is therefore given by the equation

$$T = \frac{E_3 - E_1}{k \ln(1+(g_3 A_{31}/g_1 A_{21}))} \tag{4.25}$$

which, for the parameters of a ruby laser (see Chapter 5), gives a temperature of 3300 K. The crystal in this radiation bath will receive a certain total light intensity which can be calculated by integrating Equation 2.15 over all frequency space. (This gives the familiar Stefan-Boltzmann radiation formula). On average† each square centimetre of crystal will receive about 700 watts of radiation from one direction but only about 35 watts/cm^2 will be in the absorption band, which lies from 400 to 600 nm. This must be regarded as a lower limit for the pumping intensity since it neglects several deleterious effects. In fact, continuous wave operation for a ruby laser is very difficult to achieve and almost all systems are pulsed, with the pumping source being provided by a flashtube. For this case a proper calculation would then require a solution of the time dependent equations (4.4) to (4.7). Nevertheless the static solutions give rough estimates for the required pump intensity to achieve lasing action. Of course for a more accurate approximation the absorption and output losses must be included. This can be done for the CW case by equating the small signal gain (Equation 4.20 with $W = 0$) to the distributed losses in the manner of Equation (3.68). Notice that if the pumping is provided by a flashtube then its spectral output needs to be carefully matched, if possible, to the absorption profile of the doped atoms. This is discussed with reference to a Nd-YAG laser in Chapter 5.

Another type of scheme for a three level system would involve the lasing

† The geometry is envisaged as being coaxial with the ruby rod in the centre.

transition between E_3 and E_2. In this case $B_{13} W_p$ must exceed the spontaneous decay coefficient A_{32} and the lifetime τ_{21} must be much smaller than τ_{32}.

4.2. Four level systems

A much more efficient scheme is the four level one as shown in Figure 4.2. Here atoms are transferred by some mechanism, which could of course be optical pumping, to level 4 which then decays to the upper level of the lasing transition. The lower level is now a separate level which can decay to the ground

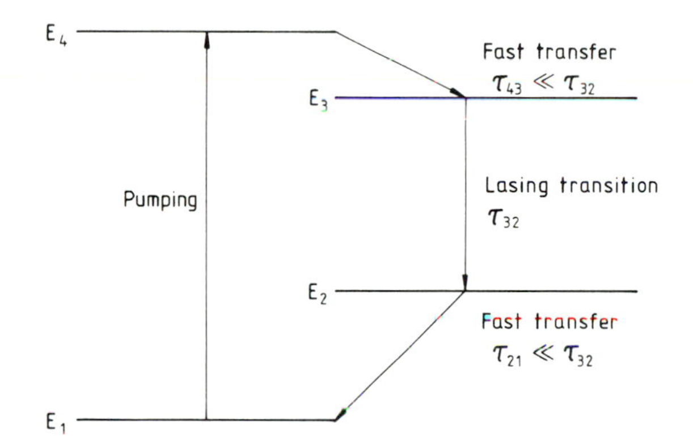

Figure 4.2. An idealized four level system.

state. As before, a set of four rate equations, together with the number density conservation equation, can be written down. Even for the steady state the solutions require a large amount of tedious algebra and the answers are cumbersome. However if the pumping is not too strong the general steady state equation (3.27) given in Chapter 3 can be used. Normally the pumping rate to level 2 is small and in these circumstances the gain coefficient for homogeneous transitions is approximately

$$\alpha(\omega) = \frac{R_3 \tau_3 \sigma_{32}(\omega)(1-(g_3/g_2)\tau_2/\tau_{32})}{1+(\sigma_{32}(\omega)I/\hbar\omega)((g_3/g_2)\tau_2+\tau_3-(g_3/g_2)(\tau_2\tau_3/\tau_{32}))} \tag{4.27}$$

where R_3 is now the total rate of transfer of atoms into level 3. Usually the decay time τ_{43} is so short that this rate can be equated to the rate at which atoms are pumped from the ground state to level 4. For weak optical pumping R_3 would be approximately

$$R_3 = n_0 B_{14} W_p(\omega_{14}) \tag{4.28}$$

for broadband illumination, or

$$R_3 = n_0 B_{14} \int_0^{\infty} h(\omega) W_p(\omega)\, d\omega \tag{4.29}$$

for narrow band illumination, where $h(\omega)$ is the absorption lineshape and $W_p(\omega)$ is the pump source output. Now for efficient operation $\tau_2 \ll \tau_{32}$ and, if level 3 decays only by a radiative transition, then the Equation (4.27) can be put in the familiar form

$$\alpha(\omega) = \frac{R_3 B_{32} f \hbar\omega}{A_{32}\eta c + B_{32} I f} \tag{4.30}$$

This applies only for homogeneously broadened transitions. For Doppler broadened transitions the small signal gain will be the same except that f will be the normalized Gaussian function centred at ω_{32}. The gain coefficient including saturation will be

$$\alpha_I(\omega) = \frac{R_3 B_{32}\, g(\omega) \hbar\omega}{A_{32}\eta c(1 + I/I_s)^{1/2}} \tag{4.31}$$

where $g(\omega)$ is the normalized Doppler profile (3.57). I_s is the saturation intensity given by the expression

$$I_s = \frac{\pi\gamma_t A_{32}\eta c}{B_{32}} \tag{4.32}$$

γ_t is the total homogeneously broadened width. When the only contribution to γ_t is from the radiative lifetime to level 2 then Equation (4.32) becomes

$$I_s = \frac{\pi A_{32}^2 \eta c}{2B_{32}} \tag{4.33}$$

4.3. *Time-dependent solutions and Q-switching*

Our analysis up to now only considered the stationary solutions to the rate equations. This is fine for an analysis of a CW laser but in a pulsed system one needs to consider the time dependence. Unfortunately the coupled equations which determine the cavity intensity cannot be solved analytically except under certain severe limitations, and one is almost always forced to use numerical techniques. In general, the solutions show damped oscillations with periodicity and damping dependent on the atomic parameters and the pumping intensity.

As an example consider a three level system as shown in Figure 4.1. If the relaxation rate is very rapid, then $n_0 = n_1 + n_2$, and the rate equations determining the population of the two levels are

$$\frac{dn_2}{dt} = n_1 W_p(t) B_{13} - \frac{n_2}{\tau_{21}} - \frac{BI(t)f(n_2 - n_1)}{c\eta} \tag{4.34}$$

and

$$\frac{dn_1}{dt} = -n_1 W_p(t) B_{13} + \frac{n_2}{\tau_{21}} + \frac{BI(t)f(n_2 - n_1)}{c\eta} \tag{4.35}$$

where, for simplicity, we assume a homogeneously broadened transition with a single mode propagation and $g_1 = g_2$. These two equations describe the most efficient pumping possible which takes place when there is an immediate transfer from level 3 to level 2. If the equations refer to a laser cavity then I is just the sum of the forward and reverse waves. Thus using our simplified travelling wave analysis† we can write

$$I = I_+ + I_- \tag{4.36}$$

and in the general plane wave case these intensities are dependent on both z and t. Subtracting (4.35) from (4.34) and using $n_0 = n_1 + n_2$ gives

$$\frac{d(\Delta n)}{dt} = n_0\left(W_p B_{13} - \frac{1}{\tau_{21}}\right) - \Delta n\left(W_p B_{12} + \frac{1}{\tau_{21}}\right) - \frac{I\Delta n 2\sigma_{21}}{\hbar\omega_{21}} \tag{4.37}$$

where

$$\Delta n = n_2 - n_1 \tag{4.38}$$

is the inversion density. A similar equation can be written for the ideal four level system (Figure 4.2) except that now the constants will be different.

Equation (4.37) cannot be solved as it stands because, of course, the intensity is time dependent, but it can also be linked to the inversion density via the gain coefficient. In Section 3.1 (Equation (3.8)) the population inversion was related to the intensity by considering the energy flow in a small volume. This equation now needs to be modified to account for the variation of intensity with time. The result for a wave travelling in one direction (I_+) is

$$\frac{\partial I_+}{\partial t} + \frac{v\,\partial I_+}{\partial z} = \alpha v I_+ - \beta v I_+ \tag{4.39}$$

† This ignores interference which causes spatial hole burning and hence the saturation characteristics will not be correctly predicted.

Here, v is the phase velocity (c/η) of the propagating beam and α is the gain coefficient which can be written in terms of the population inversion as

$$\alpha = \Delta n \sigma_{21} \tag{4.40}$$

β represents the scattering losses in the cavity in the manner of Equation (3.65). In a laser cavity there will be another equation, like (4.39), representing the reverse wave except that now the sign of the $\partial I/\partial z$ term will be opposite. This equation together with (4.37), (4.39) and (4.40) form a set of coupled equations from which, in principle, the intensity can be calculated, providing that $W_p(t)$ is known and the boundary conditions are stipulated. The general boundary conditions for plane wave propagation will involve the matching of the forward and reverse waves to the losses at the mirrors. These coupled equations are very complicated since I will depend on both z and t, but one way they can be simplified is to assume that the second term on the LHS of (4.39) is much smaller than the first. This means that, to first-order, the cavity intensity is independent of z, which will be a good approximation when the output coupling is small. The equation for the total of the forward and reverse waves ($I_+ + I_-$) is then

$$\frac{dI}{dt} = I \Delta n \sigma_{21} v - \frac{I \delta_c v}{2L_c} \tag{4.41}$$

where the diffraction, output and scattering losses have been lumped into a single distributed loss δ_c given by Equation (3.155). In the steady state, this equation is identical to (3.154). The total problem is now reduced to the two coupled equations (4.41) and (4.37). Unfortunately these are non-linear because of the product $I\Delta n$, and analytical solutions are not possible† except by numerical techniques.

Figure 4.3 shows a typical numerical solution for a ruby laser. All of these show damped oscillations whereas in practice the output of a ruby laser is often undamped. This can be predicted if extra terms are included in the coupled equations, and mechanisms have been proposed which will give rise to these terms. It should be noted that the problem has been formulated in an ideal way and several things have been neglected. For example, it assumes that a single mode‡ is propagating and the transition is completely homogeneous. In addition several practical effects such as the non-uniform temperature in the crystal lattice have been neglected. All of these need to be considered and it is not surprising that our simplified model, although it gives a plausible explanation of relaxation oscillations, is not the complete picture. In other types of laser which are not strongly pumped it is possible for the solutions to be completely damped and the intensity then rises in a smooth exponential manner to the static value.

† For small oscillations they can be solved by a perturbation method.

‡ Strictly, there can be no time dependence of the intensity if the wave is completely monochromatic. In this case the oscillations are of μs periodicity so the broadening of the line is insignificant.

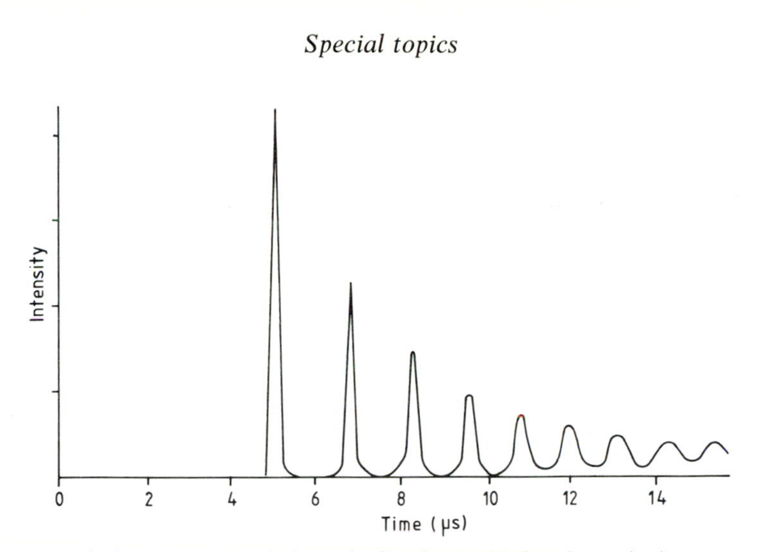

Figure 4.3. Typical computer calculations of relaxation oscillations in a ruby laser.

Relaxation oscillations in a ruby laser cause the instantaneous power output, during the spike, to increase well above that expected on a static basis. However, a much more efficient way (for any laser) to produce high power is to generate intense short pulses by means of *Q*-switching. Consider what happens when the pumping starts. The population inversion increases rapidly and coherent light is produced. If the system is critically damped the output intensity rises smoothly to the point at which the gain in a round trip is balanced by the losses. On the other hand, for an underdamped system the intensity overshoots the continuous value. In either case, if the light is allowed to propagate during the pumping, it has the effect of trying to reduce the inversion density. While the pumping is trying to increase the inversion the laser light in the cavity is acting as an effective braking mechanism preventing too large a build up. Now, if the light is not allowed to propagate during the pumping a much larger inversion can be obtained. The cavity can then be switched to its normal state, whereupon the gain coefficient starts from a very high value, and this has the effect of producing a very short intense pulse. This effect is known as *Q*-switching because the cavity *Q* value is altered from a low value during pumping to a much higher one at the start of the pulse generation. The process is essentially a means of storing energy until it is required.

The switching can be achieved in several ways. One way, to produce relatively slow switching, is to rotate one of the mirrors. The pumping is synchronized to take place just before the mirrors pass through the parallel position where the giant pulse is generated. A more rapid switching can be brought about using various electro-optic shutters which block out one of the mirrors. As soon as the pumping is complete and the inversion is high the shutter is opened electronically. Another technique is to use a dye which is bleached as the radiation intensity increases. The dye behaves essentially like a two level system and becomes transparent as soon as its inversion density is zero. (Note that Δn starts from a negative value in this case!) This does not have a complete on/off nature so the stored energy is less but this is balanced by the fact that the switching is very rapid and the system is easy to implement.

The analysis of Q-switching again uses the coupled equations (4.37) and (4.41). These will be valid as long as we can neglect the spatial variation of intensity. In Q-switching the second term on the RHS of (4.41) represents the losses or damping. When Q is small the losses δ_c are large, whereas when Q is large, during the pulse formation, δ_c is small. A complete analysis would require a numerical solution to these equations making δ_c time dependent. A simplified model can be obtained by assuming that the switching is represented by a step function. Further, the pumping is represented by a square pulse which stops at the moment of switching. If the Q is very low during the pumping, then the inversion density can be obtained from 4.37 by ignoring the last term. The inversion is then given by the equation

$$\Delta n = \frac{n_0(W_p B_{13} - 1/\tau_{21})}{(W_p B_{13} + 1/\tau_{21})}(1 - 2\exp(-W_p B_{13} t - t/\tau_{21})) \tag{4.42}$$

and since $1/\tau_{21}$ is much smaller than $W_p B_{13}$ this is approximately

$$\Delta n = n_0(1 - 2\exp(-W_p B_{13} t)) \tag{4.43}$$

After a long time compared with $1/W_p B_{13}$, all of the atoms have been pumped into the higher state. Alternatively, the losses may not be so large. In this case the inversion can be obtained by assuming that, after a long time (longer than $1/W_p B_{13}$), dI/dt and $d\Delta n/dt$ are zero. The intensity is then obtained in the same way as for CW operation (see Section 3.7) and this can be substituted into (4.37) to give the inversion density.

Now consider what happens after the switching. If the pumping is now zero the two coupled equations become

$$\frac{dI}{dt} = I\Delta n \sigma_{21} v - I\Gamma \tag{4.44}$$

and

$$\frac{d(\Delta n)}{dt} = -I\Delta n \frac{2\sigma_{21} v}{\hbar\omega_{21}} \tag{4.45}$$

where Γ now represents the damping and is explicitly $\delta_c v/2L$. These two equations are often written in terms of the total number of cavity photons in the particular mode. According to Equation (2.100) the intensity is related to the photon density ϕ by the equation

$$I = \phi\hbar\omega v \tag{4.46}$$

where v is now the phase velocity. The total number of photons in the cavity, Φ, is ϕV where V is the volume† so (4.44) and (4.45) can be written

$$\frac{d\Phi}{dt} = b_{21}\Phi\Delta N - \Gamma\Phi \tag{4.47}$$

and

$$\frac{d(\Delta N)}{dt} = -2b_{21}\Phi\Delta N \tag{4.48}$$

where b_{21} is $\sigma_{21}\, v/V$. This is just $B_{21}\, f(\omega)\hbar\omega/\eta^2 V$.

$\Delta N = N_2 - N_1$ now represents the total inversion in the cavity. Equations (4.47) and (4.48) are only applicable for single mode propagation when the transition is homogeneously broadened. If there is no damping ($\Gamma = 0$) the two equations have solutions

$$\Delta N = \frac{\Delta N_0}{2}\left[1 - \tanh\left(\frac{b_{21}\,\Delta N_0(t - t_0)}{2}\right)\right] \tag{4.49}$$

and

$$\Phi = \frac{\Delta N_0}{4}\left[1 + \tanh\left(\frac{b_{21}\,\Delta N_0(t - t_0)}{2}\right)\right] \tag{4.50}$$

Here ΔN_0 is the inversion just before switching, and t_0 is the time when this has reduced to $\Delta N_0/2$. These equations give a good approximation to the rising shape of the pulse when the damping is much less than $\Delta N_0\, b_{21}$. Notice that these equations imply that, if Φ is zero, at the beginning, then there is an infinite time to the $\Delta N_0/2$ point. In other words, the solutions only have a meaning if there is a finite value of Φ before the switch. However, Equation (4.47) is not strictly correct and there should be a very small extra term, $N_2\, b_{21}$, which will allow the emission to begin. For very long times the number of photons in the cavity tends to $\Delta N_0/2$ when the inversion goes to zero. At this point the photons are in equilibrium with the excited atoms. When the damping is included Φ turns over before reaching the maximum and decreases, as shown in Figure 4.4.

The important parameters of the pulse shape for weak damping can be

† Assuming for simplicity that the active medium fills the cavity so $L_c = L$.

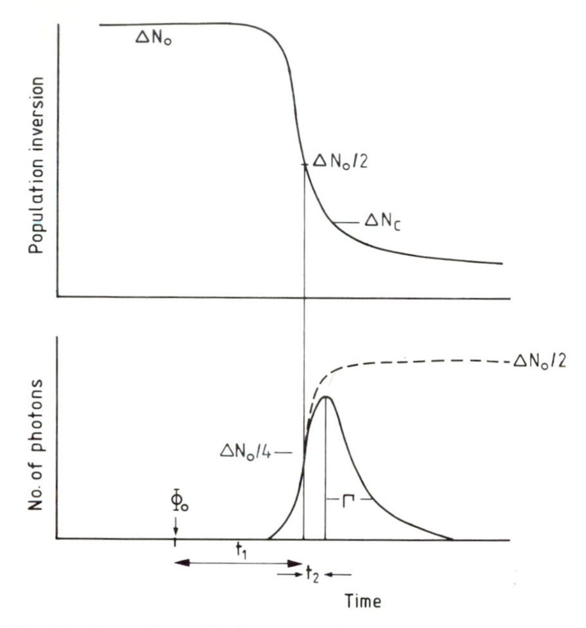

Figure 4.4 Total number of photons and the population inversion for a Q-switched laser. Approximate values of t_1 and t_2 are given for the condition $\Delta N_0/\Delta N_c \gg 1$.

estimated as follows. Firstly, the time it takes for the photon number to go from an initial value Φ_0 to $\Delta N_0/4$ (the half point where ΔN is $\Delta N_0/2$), can be found directly from (4.50). This is

$$t_1 = \left(\frac{-2}{b_{21}\,\Delta N_0}\right)\tanh^{-1}\left(\frac{4\Phi_0}{\Delta N_0} - 1\right) \simeq \left(\frac{1}{b_{21}\,\Delta N_0}\right)\ln\left(\frac{\Delta N_0}{2\Phi_0}\right) \tag{4.51}$$

where the approximate expansion $\tanh(-x) = 2\exp(-2x) - 1$ has been used. Now, from Equation (4.47), it is clear that the peak of the output ($d\Phi/dt = 0$) occurs at the point where the inversion density is that required to give CW operation in the event of continuous pumping. If this inversion is ΔN_c then, from (4.47),

$$\Delta N_c = \frac{\Gamma}{b_{21}} \tag{4.52}$$

and, according to (4.49), the time to go from t_0 to this point is

$$t_2 = \left(\frac{2}{b_{21}\,\Delta N_0}\right)\tanh^{-1}\left(1 - \frac{2\Delta N_c}{\Delta N_0}\right) \simeq \left(\frac{1}{b_{21}\,\Delta N_0}\right)\ln\left(\frac{\Delta N_0}{\Delta N_c}\right) \tag{4.53}$$

The maximum value of Φ at this time is, from (4.50),

$$\Phi_m = \frac{(\Delta N_0 - \Delta N_c)}{2} \tag{4.54}$$

After the peak the pulse decays with an approximate shape

$$\Phi = \Phi_m \exp(-\Gamma t) \tag{4.55}$$

4.4. *Mode locking*

Q-switching can produce pulses with durations as short as 10^{-8} s, and during this time almost all the energy stored in the medium is released. (For example, a small ruby crystal might store about 10 J. With suitable amplification, systems can be designed to produce pulses with peak powers in excess of 10^{12} watts. These systems are often used in fusion or plasma research, where very high powers are required.) Even shorter pulses, less than one picosecond, can be produced by mode locking. These short pulses can be used to study, or probe, extremely rapid phenomena such as photochemical reactions. In some cases, mode locking and Q-switching can be combined to produce phenomenal power levels.

The concepts of mode locking are relatively simple although a full mathematical treatment is rather involved. Here we shall be content to give a much simplified picture. Consider what happens in a laser which is oscillating at several different mode frequencies. It is envisaged that the gain profile is very much greater than the mode spacing. For a simple analysis we take a linearly polarized plane wave to represent the positive going wave. The electric field of this wave in the cavity can be written

$$E = \sum_n A_n \sin(\omega_n \eta z/c - \omega_n t + \delta_n) \tag{4.56}$$

where ω_n are the mode frequencies and A_n are the amplitudes. The total electric field in the cavity will have an added term, representing the reverse wave, which will have the phases altered by π and the sign of $\omega_n t$ reversed. The output wave will have the same form as (4.56) except that the amplitudes A_n will be reduced. Now, in general, the phases, δ_n, of each mode, are random so that the intensity averaged over one cycle of the field will be constant. However, let us suppose that the phases can be locked together so that δ_n is constant. To make matters simpler it is also assumed that the amplitudes are constant, although an analysis is possible with the values of A_n modulated with a Gaussian profile. (This would be the case for inhomogeneous broadening). The total electric field is then

$$E = A_0\, Re\left[\exp(i\omega_0 + \delta) \sum_n \exp(in\Delta\omega\phi)\right] \tag{4.57}$$

with

$$\phi = \frac{\eta z}{c} - t \tag{4.58}$$

and the mode frequencies, ω_n, are related to the lowest mode, ω_0, by the equation

$$\omega = \omega_0 + n\Delta\omega \tag{4.59}$$

where $\Delta\omega$ is the mode spacing, which is just

$$\Delta\omega = \frac{\pi c}{L\eta} \tag{4.60}$$

The summation in Equation (4.57) is

$$\sum \exp(in\Delta\omega\phi) = \frac{1 - \exp(iN\phi\Delta\omega)}{1 - \exp(i\phi\Delta\omega)} \tag{4.61}$$

where N is the total number of modes. Using the relation $1 - \exp(i\delta) = -\exp(i\delta/2) \,.\, 2i \sin \delta/2$, Equation (4.57) is finally simplified to

$$E = A_0 \sin \omega_m \phi \frac{\sin(N\Delta\omega\phi/2)}{\sin(\Delta\omega\phi/2)} \tag{4.62}$$

where ω_m is the average mode frequency given by the expression

$$\omega_m = \omega_0 + (N-1)\frac{\Delta\omega}{2} \tag{4.63}$$

The wave is therefore the same as a single mode at ω_m except that now it is modulated. The output intensity will have the form

$$I = \left[\frac{\sin(N\Delta\omega\phi/2)}{\sin(\Delta\omega\phi/2)}\right]^2 \tag{4.64}$$

where I_0 is the single mode intensity. Figure 4.5 plots the intensity for $N=6$. Considering the output as a function of time, the intensity is increased by a factor

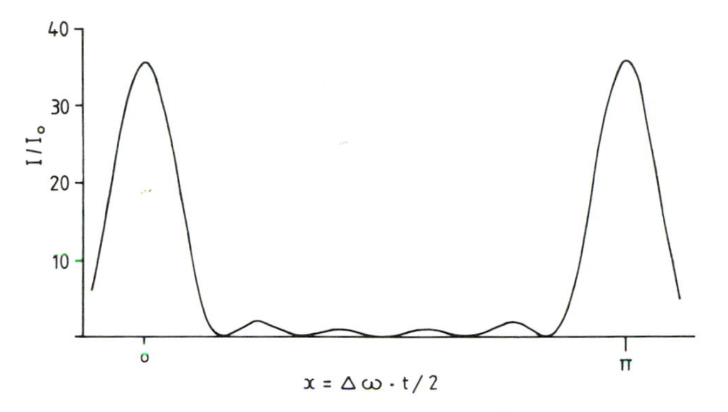

Figure 4.5. The intensity of six equal amplitude modes when their phases are the same.

of N^2, when $\Delta\omega t/2 = k\pi$, which means that the spacing between pulses is

$$\Delta t = \frac{1}{\Delta \nu} = \frac{2L\eta}{c} \tag{4.65}$$

corresponding to the time it takes for a single pulse to make a complete round trip between the mirrors. For large N, the denominator in (4.64) is much more slowly varying than the numerator, so the width of the pulse at the base is

$$\delta t \simeq \frac{1}{\Delta\omega N} \tag{4.66}$$

but since the number of cavity modes is roughly the full width of the gain profile, $2\gamma_I$, divided by the mode spacing, then (4.66) is simply

$$\delta t \simeq \frac{1}{2\gamma_I} \tag{4.67}$$

For example, a gas laser might have a typical broadened profile 2 GHz wide so the minimum pulse width is about 0·5 ns. In doped insulator lasers $2\gamma_I$ might be as large as 3000 GHz and, hence, pulses of picosecond duration are possible.

Essentially there are two methods of locking the phases together. In active mode locking an electro-optic device† is used to amplitude modulate the light in the cavity. The frequency of modulation is chosen to match the mode spacing. The modulation of any mode produces sidebands at exactly the positions of the other modes, which will eventually lock the phases together. To imagine how the process begins, assume that only the single centre mode has sufficient gain to oscillate, but the overall gain is gradually increasing. The sidebands of this centre

† Description in Chapter 5.

mode will initiate the start of lasing action in the adjacent modes which will, in turn, initiate oscillations in further adjacent modes. In this way all the modes ultimately become locked. Another way to look at this process is to image what happens to the photon burst as it travels around the cavity. The amplitude modulation is represented by a shutter which operates at one of the mirrors. If this allows a maximum transmission every $2L\eta/c$ seconds then parts of the pulse which take less or more than this are attenuated. There is therefore a tendency to compress all the beam energy into a narrow spike.

In passive mode locking a dye solution is used in the same way as Q-switching. If the relaxation in the dye is rapid then it can respond almost instantaneously to intensity changes in the cavity. A fluctuation, or intensity pulse, will suffer less attenuation because the dye will become more bleached. There is therefore a strong tendency to compress the energy into a narrow spike, which is achieved when the modes are locked.

4.5. *Spatial hole burning*

So far the analysis of the laser output and saturation characteristics has been based on a simplified intensity analysis. This is correct for a travelling wave amplifier but is not true for a standing wave oscillator. This can be seen quite simply by calculating the total electric field in the cavity for single mode operation. Making the approximation of a linearly polarized plane wave with small output losses the electric field is

$$E = E_0(\sin(kz-\omega t+\delta)+\sin(kz+\omega t+\delta+\pi)) \tag{4.68}$$

Here the reverse wave is altered in phase by π on reflection. Equation (4.68) is simply reduced to

$$E = 2E_0\cos(\omega_n t)\sin(kz+\delta) \tag{4.69}$$

and for perfectly reflecting metallic mirrors, δ would be zero if z is measured from the surface of either mirror. Now the probability of absorption or stimulated emission is determined by E^2 (see Equation 2.118), so, in the standing wave pattern governed by (4.69), there will be regions of the lasing medium which will not be used effectively. The gain of the system for single mode propagation will be lower than calculated because the medium will not saturate uniformly. This is in contrast to a travelling wave where the intensity averaged over one field cycle (which is proportional to E^2) is uniform.

If many modes are present each one will burn a different pattern in the medium. The exact nature of these will depend on the length and position of the lasing medium in relation to the total cavity. Any alteration in cavity length caused by acoustic vibration, or changes in refractive index, will alter the pattern and give rise to a coupling between the different modes. The competition between modes can be a problem in a system designed to operate at a single frequency. If the amplifying medium is much shorter than the cavity length, as, for example, in a dye laser, then the tendency to change mode (mode-hopping) can be minimized by placing the dye close to one of the mirrors. It is then fairly easy to see that there needs to be a much larger shift in wavelength, or mode number, before the standing wave pattern will affect different parts of the dye, and hence compete for gain. Ideally, this shift should be larger than the gain profile of the laser system, which in a dye laser can be achieved using intra-cavity etalons. Another way to avoid spatial hole burning effects is to use a travelling wave oscillator, or ring laser.

4.6. Linewidth of a single mode laser

In many spectroscopic applications it is important to know the linewidth of the laser output. Hitherto it has been assumed that in single mode operation the output is a delta function, but clearly this cannot be true. The linewidth in all systems is determined by practical considerations such as the stability of the mirror separation and changes in the optical path length caused by alterations of the refractive index. For example, the mode spacing in a 1 m cavity is 150 MHz and the separation of the mirrors must be held to much better than $\lambda/2$ ($\simeq 0.3\ \mu$m in the visible) for true single mode operation. This can be achieved relatively easily, particularly if the optical path length is adjusted in response to alterations of the output wavelength. These actively stabilized systems have have linewidths of 1 MHz or less for visible radiation.

It is instructive to ask what the maximum attainable linewidth would be in an ideal system where the optical path length is fixed. The linewidth then depends on the noise generation within the amplifier of the oscillator. For example, amplitude modulations of the cavity power will cause corresponding frequency broadening in the output. These amplitude modulations can be thought of as statistical fluctuations in the number of photons in any one cavity mode. Another source of broadening will be caused by the spontaneous emission which happens to lie within the particular mode of interest. However, because there are many modes available, and also because the total intensity of this radiation is small in comparison to the stimulated emission, it can be safely ignored. In fact, the largest contribution to the linewidth comes from phase fluctuations due to amplified spontaneous emission. Any photon spontaneously emitted into the cavity mode will be amplified and the phase of the resulting contribution will not necessarily be the same as the propagating beam. This will not cause an amplitude fluctuation because of gain saturation, but, rather, a diffusion or spreading of the overall phase which can be related to a line broadening. A detailed analysis of the noise contributions gives the Schawlow-Townes equation

$$\Delta v = \frac{\pi h v \Delta v_c^2}{p} \tag{4.70}$$

for the output linewidth, Δv, in terms of the passive cavity response Δv_c (Equation 3.95). p is the output power and v is the frequency. As an example, consider the He–Ne laser described at the end of Section 3.7. The Q value is of the order 10^9 so Δv_c will be approximately 0·6 MHz. For an output power of 1 mW the linewidth is then 7×10^{-5} cycles per second.

Extremely expensive systems have been constructed with linewidths of a few Hz, but, clearly, practical considerations play the dominant role in determining the frequency broadening of the output.

Bibliography

Birnbaum, G., 1964, *Optical Masers* (New York: Academic Press).
Maitland, A. and Dunn, M.H., 1969, *Laser Physics* (Amsterdam: North Holland Publishing Company).
Shimoda, K., 1948, *Introduction to Laser Physics* (Berlin: Springer-Verlag).
Verdeyen, J.T., 1981, *Laser Electronics* (New Jersey: Prentice Hall).

CHAPTER 5

solid and liquid lasers

5.0. Introduction

The remainder of this book is devoted to descriptions of a wide variety of laser systems. Because of the limitations of space, the explanations of each type will necessarily be somewhat brief, although in all cases the mode of operation and important characteristics of the output light are given. These characteristics are most important when deciding whether a specific laser is suitable for one particular application or another. Often this will depend on the additional elements in the system, either within or external to the cavity, and, for this reason, examples of how the laser output can be modified are given in a number of cases. Also, some important topics which have not been described previously, for example, ring lasers, will be included.

Lasers can be classified in a number of different ways, but perhaps the most useful is to separate them according to the physical nature of the amplifying medium; the major categories being doped insulator, gas, semiconductor or liquid. Within these areas one can distinguish separate types of laser which are of a different nature and require separate descriptions. For example, chemical lasers are a type of gas laser but the nature of their operation is quite different from the majority of other gas lasers. Also, it should be borne in mind that, in many cases, the concentration of atoms or molecules which take part in the amplification (n_0, n_1, n_2, etc of our previous analysis) is much less than the host medium, so in one sense this classification is concerned with the changes of level structure, relaxation processes, or methods of pumping, which are brought about by the different environments of the active atoms† (ions or molecules). This does not necessarily apply in many gas lasers, where, because of the low pressure, the concentration of atoms is low anyhow. In the case of liquid lasers we shall only discuss dye laser systems, particularly tunable systems for spectroscopy.

† In semiconductors the electrons cannot be said to belong to any one atom or ion.

5.1. Doped insulator lasers

Doped insulator lasers are one of the most prolific types, and are widely used in systems where high power short duration pulses of light are required, for example, in plasma or fusion research. (For very high powers, amplifiers are used.) In spectroscopy, they can be used as a primary pump source for a dye laser if the output is frequency doubled. The principle of these lasers depends on the fluorescent properties of various crystalline (or glassy) solids when doped with certain elements (rare-earth or transition elements). The dopants usually become part of the lattice, rather than residing at interstial sites, and, as such, they are essentially in ionic form. It is the excited energy levels of these dopant ions which provide the lasing action. However, these energy levels are considerably modified from the bare ion, because of the high electric fields within the lattice, and, furthermore, the levels will depend on the specific host material.

Optical pumping is used to populate excited states of the ion which can then decay rapidly to other excited states by photon emission or relaxation processes. In three level systems, like ruby, the lasing transition is between these excited states and the ground state whereas in a four level system, like Nd-YAG, the transition is to a low lying excited state. The crystal has two major influences apart from the obvious one of containing the dopants. Firstly, it causes a general broadening of the levels which is important for efficient pumping with broadband illumination, particularly for three level systems. Secondly, the excited states have the possibility of decay by relaxation processes where the energy is transferred directly to the lattice by phonon emission rather than photon emission. This process can be much more rapid than electromagnetic decay, particularly when the transition energy is small or is partly forbidden by selection rules.

In this section we will deal specifically with the ruby laser and the Nd-YAG system, the latter being the most widely used of all these types.

Ruby laser

The ruby laser occupies a rather special place in the history of the laser, since it was the first system to operate in the visible spectrum, all previous systems being masers.

Ruby is a crystalline form of Al_2O_3 (corundum) containing between 0·05% and 0·5% by weight of Cr_2O_3. At the lower concentration it is pale pink whilst at higher concentrations it becomes deep red. The chromium resides in lattice in the form of Cr^{+++} ions ($1{\cdot}6 \times 10^{19}$ Cr^{+++} ions per cm^3 for the lowest concentration which is the usual form in lasers), and these ions endow the crystal with certain fluorescent properties, which make it a suitable amplifying medium for use in an optically pumped laser. Now the chromium ion has three electrons in the $1d$ shell with the lowest levels belonging to a ^4F and a ^2G term. Within the crystal there is a considerable modification of the energy levels brought about by the high electric fields associated with ionic bonding. This change is more complicated than a simple Stark splitting because the crystal has a 3-fold axis of symmetry. The ^4F term is split into 3 levels usually designated by the symbols $^4F_1$, $^4F_2$, and $^4A_2$. Here, only the spin multiplicity is retained; F_1, F_2, and A_2 are used to designate the different matrix representations of the octahedral group, and should not be confused with the L or J values associated with the infinitesimal rotation group. The total multiplicity of the $^4F_1$, $^4F_2$ and $^1A_2$ levels are 12, 12 and 4 respectively. In a similar way the ^2G term is split into 4 levels designated by the symbols $^2A_1$, $^2F_1$, $^2F_2$ and ^2E with total multiplicities of 2, 6, 6 and 4 respectively. In addition, the ^2E level is not a single level but a pair of closely spaced levels approximately 970 GHz apart. These are usually noted by the symbols $2A$ and E. Because the chromium ions vibrate in the lattice all of these energy levels are broadened to a lesser or greater extent. In fact, the $^4F_1$ and $^4F_2$ levels are approximately 60 nm wide giving rise to a broad absorption band between 350 and 600 nm. The widths of the other levels are much smaller, for example, the total width of E is about 0.7 nm at 60°C. Furthermore, both the widths and the energy spacings are temperature dependent because the vibration amplitudes and lattice spacings change with temperature.

Figure 5.1 shows the pertinent energy levels for a ruby laser, which is essentially a three level system. Optical pumping is used to populate the levels $^4F_1$ and $^4F_2$ which can then undergo rapid relaxation (phonon emission) to the narrow ^2E levels. This relaxation is very rapid ($\tau_{32} \sim 5 \times 10^{-8}$ s) and it is reasonable to assume that almost all the atoms which absorb a photon are transferred to ^2E because, even when the pumping is intense, the rate of stimulated emission of $^4F_1$ and $^4F_2$ to the ground state is still much smaller than

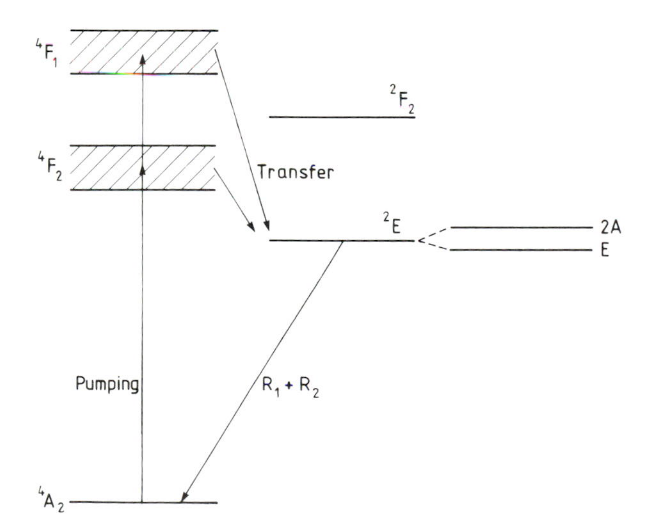

Figure 5.1. Energy levels for a ruby laser.

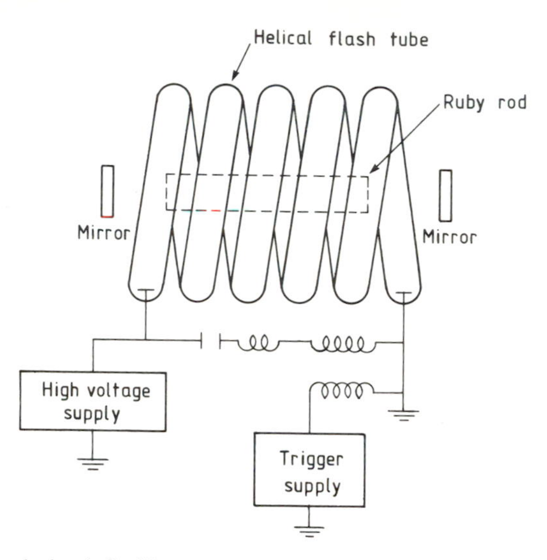

Figure 5.2. A simple flashlamp pumped ruby laser.

S_{32}. The transition from either E or $2A$ to the ground state, known as R_1 and R_2 respectively, can be made to lase. At room temperature the lifetime for R_1 fluorescence is 3 ms, and the wavelength is 694·3 nm. E and $2A$ are strongly coupled by relaxation of the crystal, a process which is temperature sensitive since it depends on the availability of lattice phonons. The populations of the two levels are related by the equation $n(E) = C(T)n(2A)$, where $C(T) = 1$ at room temperature. Both E and $2A$ can decay by phonon-assisted processes, whereby only part of the energy is in the form of a photon, the remainder being transferred directly to the lattice as phonons. At 0°C, the fraction of decays from E, which contribute to the R_1 line, is 0·55.

Two practical systems for pulsed ruby lasers are shown in Figures 5.2 and 5.3. In both cases the optical pumping is provided by a flash lamp. This is a glass

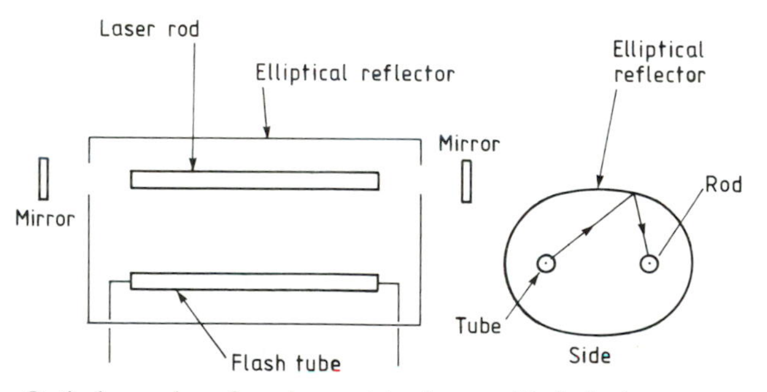

Figure 5.3. Optical pumping of a ruby crystal using an elliptical mirror.

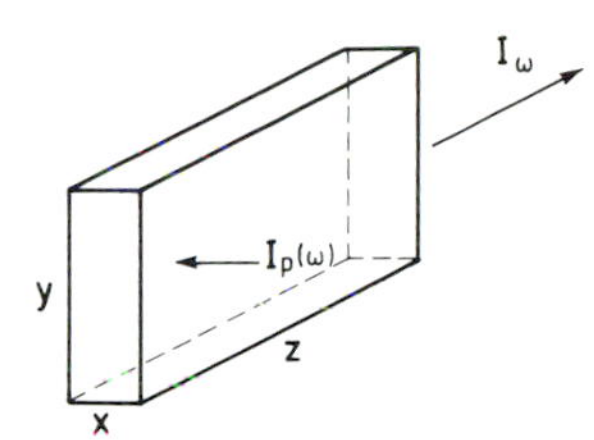

Figure 5.4. Geometry for optical pumping with unidirectional light.

tube filled with an inert gas, usually xenon for a ruby laser, through which is passed a large electric current at high voltage. Typically, a single pulse might be produced by discharging a capacitor of a few hundred joules of stored energy in about 200 μs. The storage capacitor is usually at a voltage of a few kV but a much higher voltage is required to initiate the breakdown (10 kV), and one way this can be accomplished is shown in Figure 5.2. The output light is distributed from the UV all the way to the infra-red in an irregular manner which can be altered in varying the voltage across the tube. Typically about 30% of the optical emission lies in the absorption band for ruby.

Now, the optical pumping of a ruby laser is extremely complicated for any practical geometry even if we make the simplifications of Section 4.1. For example, Figure 5.4 shows a simplified geometry in which the pumping light is coming from a single direction. The time dependent rate equations for the population of levels 1 and 2 are

$$\frac{\partial n_2}{\partial t}(\boldsymbol{r}, t) = n_1(\boldsymbol{r}, t) \int \left(\frac{I_p(\boldsymbol{r}, t, \omega)}{\hbar\omega} \right) \sigma_{13}(\omega)\, d\omega - \Delta n(\boldsymbol{r}, t) \left(\frac{I_\omega(\boldsymbol{r}, t)}{\hbar\omega} \right) \sigma_{21}(\omega) \tag{5.1}$$

and

$$\frac{\partial n_1}{\partial t}(\boldsymbol{r}, t) = -n_1(\boldsymbol{r}, t) \int \left(\frac{I_p(\boldsymbol{r}, t, \omega)}{\hbar\omega} \right) \sigma_{13}(\omega)\, d\omega + \Delta n(\boldsymbol{r}, t) \left(\frac{I_\omega(\boldsymbol{r}, t)}{\hbar\omega} \right) \sigma_{21}(\omega) \tag{5.2}$$

where I_p is the spectral intensity of the pump, and we are assuming a single mode propagating with intensity I_ω. These equations are the same as (4.34) and (4.35) except for the spatial dependence of the pumping. At room temperatures and above the problem is essentially homogeneous in nature and can be formulated by these two equations, whereas at low temperatures the ions 'freeze' out at different lattice positions giving rise to inhomogeneous effects. The equations would then include the lattice position in the same manner as (3.41) and (3.42) contained the velocity. (Note that the shift in frequency as a function of position would need to be known before f_2 (Equation (3.43)) could be formulated.) As well as these two equations, we also need to construct two sets of equations, like (4.39) and (4.40), for both the pump and the output light. (The pump equations would involve terms like $\partial I_p/\partial x$ and $\partial I_p/\partial t$). Clearly, the problem is very difficult and certainly outside the scope of this book.

To simplify the problem, assume that the system is Q-switched so that the σ_{21} term of (5.1) and (5.2) can be ignored at first. Now, we wish to just study the absorption of pump light, and make some approximate calculations for the intensity and duration of light, required to produce a uniform population inversion. Firstly, the absorption coefficient for a parallel pumping beam is

$$\alpha(\omega, x, t) = n_1(x, t)\sigma_{13}(\omega) \tag{5.3}$$

Figure 5.5 shows how the cross section, σ_{13}, varies as a function of the wavelength of light. For a dopant density of $1{\cdot}6 \times 10^{19}$ atoms/cm^3, the value of $\alpha(t = 0)$, at the peak of the absorption cross section ($\lambda \sim 400$ nm), is $6{\cdot}4\,\text{cm}^{-1}$. This means that the spectral intensity at this wavelength is reduced by a factor of $1/e$ in a distance of about 0·15 cm. Of course, for frequencies away from the maximum, the pump light penetrates further into the crystal. Now this does not mean that the crystal should have dimensions which are smaller than the maximum value of $1/\sigma$,

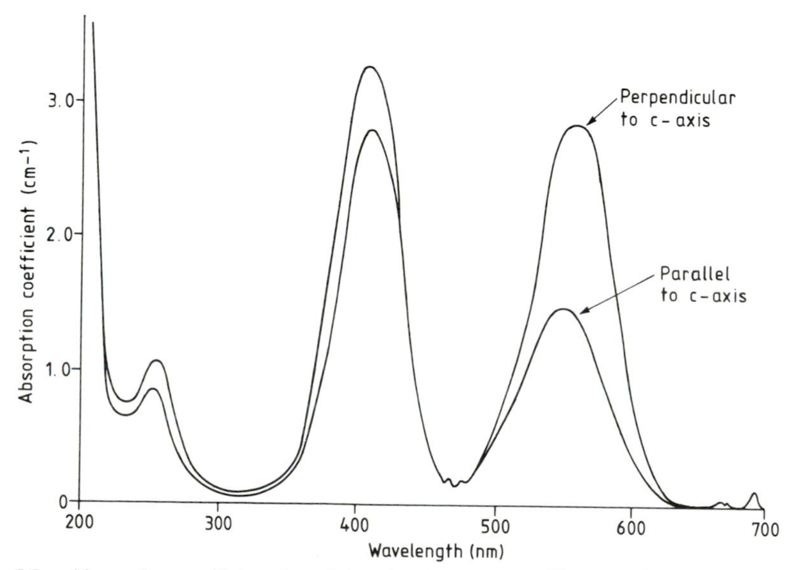

Figure 5.5. Absorption coefficient for pink ruby ($n_0 = 1{\cdot}6 \times 10^{19}$ per cm^3). The absorption is different for the incident light perpendicular and parallel to the c-axis. (From D.C. Cronemeyer, *J. Opt. Soc. Am.* **56** (1966) 1703.)

because the first layers of the crystal become partly bleached. What is necessary is for there to be a sufficient number of photons in the required wavelength range to produce a fairly uniform distribution with everywhere $n_2 > (g_2/g_1)n_1$. The total energy per unit volume, taken up by the crystal is (ignoring spatial effects which would alter σ and ε throughout the bulk of material)

$$E_a = n_2(t)\frac{\int \hbar\omega\, \sigma_{13}(\omega)\varepsilon(\omega)\, d\omega}{\int \sigma_{13}(\omega)\varepsilon(\omega)\, d\omega} \tag{5.4}$$

where $\varepsilon(\omega)\, d(\omega)$ is the fraction of the total intensity in the finite frequency interval from ω to $\omega + d\omega$. Assuming a reasonable value of $n_2 = \frac{3}{4}n_0$, this is very approximately

$$E_a \simeq \tfrac{3}{4}n_0\, \hbar\bar{\omega} \tag{5.5}$$

with $\hbar\bar{\omega}$ being the average photon energy in the absorption band. Since this goes from 350 to 600 nm, $\bar{\lambda} = 475$ nm and $\hbar\omega = 2{\cdot}6$ eV, giving $E_a = 5\ \mathrm{J/cm^3}$. Of course to produce this energy in the absorption band the output energy of the flashtube will need to be considerably more. The fraction of light in the absorption band is just

$$f_1 = \frac{\int \sigma_{13}(\omega)\varepsilon(\omega)\, d\omega}{\int \sigma_{13}(\omega)\, d\omega} \simeq \frac{\Delta\lambda_a}{\Delta\lambda_p} \tag{5.6}$$

Here it is assumed that the absorption cross section is uniform over a range $\Delta\lambda_a$, and the pump output is constant in the interval of width, $\Delta\lambda_p$, and zero elsewhere. f is then approximately 280/850, or 0·29, so the light output of the tube needs to be 17 J per cm^3 of the crystal. This is not the energy stored in the capacitor shown in Figure 5.2 because some of the energy is lost as heat. Typically, only about 60% of the total energy is in the form of light with $\Delta\lambda_p = 850$ nm. A total stored energy of about 140 J is thus required to produce $n_2 = \frac{3}{4}n_0$ in a crystal 10 cm long and 0·5 cm^2 area.

The efficiency of a ruby laser is rather poor, particularly because the pumping has to put at least half of the ions into the excited state before any output is possible, so already in our example, $2/3E_a$ is effectively wasted. An estimate of the efficiency for a Q-switched system can be made as follows. Firstly, make the simplifying assumption that the two levels, $2A$ and E, are combined, so $g_1 = g_2$, and the inversion density is $n_2 - n_1$. The maximum total output energy in the form of photons per unit volume of crystal, is

$$E_0 = \frac{\Delta n_0}{2}\hbar\omega_0 \tag{5.7}$$

where ω_0 is the lasing frequency and Δn_0 is the initial inversion density. Using the previous equations (5.4) and (5.6), the efficiency is

$$\frac{E_0}{E_T} = \left(\frac{\Delta n_0}{2n_2}\right) f_1 \, f_2 \frac{\int \sigma(\omega)\varepsilon(\omega)\,d\omega}{\int \hbar\omega\sigma(\omega)\varepsilon(\omega)\,d\omega} \hbar\omega_0 \tag{5.8}$$

Here f_1 is given by (5.6) and f_2 is the fraction of the flashtube output energy in the form of electromagnetic radiation. (Excluding any radiation, UV or infra-red, absorbed by the glass enclosure of the flashtube.) Equation (5.8) is very approximately

$$\frac{E_0}{E_T} \simeq \left(\frac{\Delta n_0}{2n_2}\right) \frac{f_1 \, f_2 \, \hbar\omega_0}{\hbar\bar{\omega}} \tag{5.9}$$

and putting $n_2 = (3/4)n_0$, as before, and suitable values for the other quantities, gives $E_0/E_T \simeq 0{\cdot}04$. This should be considered as an overestimate for several reasons, not the least being the approximation to evaluate the integrals. Also we have neglected any spatial effects in the absorption of pump light and assumed that all the cavity photons can be extracted with no absorption losses. Similarly, we have ignored unwanted absorption of the pump light, both within the crystal and at the reflectors.

Notice in the previous calculation we also ignored the fact that there are two possible laser transitions. If the laser is operating only on the lower line (R_1) then the coupling between the two states needs to be considered. When the pulse length is long compared with the relaxation time (1 ns), the lower level, E, can be repopulated continuously, from $2A$. For a no loss system (impossible, of course!), the oscillations cease when $n_E = n_1/2$, since $g_1 = 2g_E$. At this point, $n_{2A} = n_E = n_0/4$ and $n_1 = n_0/2$. The maximum number of photons is therefore the same as previously. On the other hand, if the pulse is very rapid (mode locked) then only the initial population of E needs to be considered. If n_2 is the total number of ions pumped into the upper levels, then lasing is extinguished when $n_E = 1/3(n_0 - n_2/2)$ and $n_A = n_2/2$, so the total number of photons extracted is $2/3n_2 - 1/3n_0$. This is to be compared with the larger value, $n_2 - n_0/2$, for long pulses. This cross relaxation can be modelled using coupled rate equations.

Now let us look at the stimulated emission and calculate the approximate shape of the output pulse using the analysis of Section 4.3. The most important parameter is the cross section, σ_{21}, which is directly proportional to the gain coefficient (see Equation (4.40)). The cross section, for monochromatic light of angular frequency ω, is

$$\sigma_{21} = \frac{B_{21} \, f(\omega)\hbar\omega}{\eta c} = \frac{B_{21} \, \bar{f}\hbar\omega}{\eta c\pi\gamma_t} \tag{5.10}$$

where γ_t is the half width of the homogeneously broadened transition and $\bar{f}$ is the

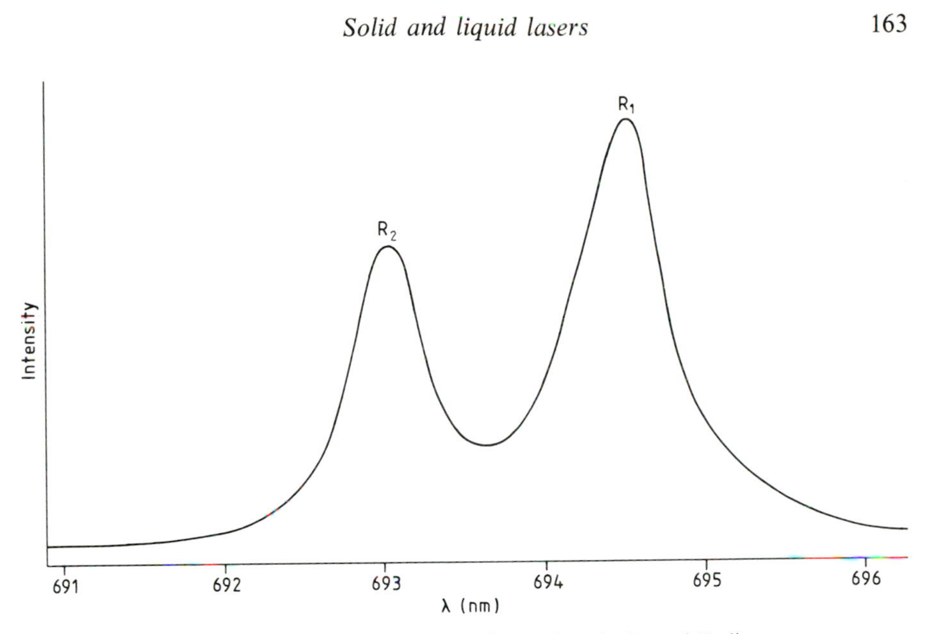

Figure 5.6. Fluorescent spectrum for ruby at 60°C, showing the R_1 and R_2 lines.

Lorenzian (3.34). It is much more convenient to write this cross section in terms of the A coefficient or lifetime, using (2.142) and (2.24)†. However, the relation between A and B was derived in a vacuum and needs modifying to account for the refractive index of the medium. This is done by introducing an η in the numerator of the RHS of (2.24). The cross section then becomes

$$\sigma_{21} = \frac{\pi c^2 A_{21} \bar{f}}{\omega^2 \eta^2 \gamma_t} = \frac{\pi c^2 \bar{f}}{\omega^2 \eta^2 \gamma_t \tau_{21}} = \frac{\pi \lambda\!\!\!^{-2} \bar{f}}{\gamma_t \tau_{21}} \tag{5.11}$$

where τ_{21} is the partial lifetime for spontaneous emission. If the broadening is due entirely to spontaneous emission, then (5.11) reduces to $\sigma = 2\pi\lambda\!\!\!^{-2}$ at the peak of the distribution. In this case both σ_{21} and σ_{12}, the resonant absorption cross section, are many orders of magnitude larger than the geometrical cross section of the atom. This fact is exploited in many areas of ultrasensitive laser spectroscopy. However, for ruby, $2\gamma_t \gg 1/\tau$, since the broadening is due to other factors. Figure 5.6 shows the spectral distribution for ruby fluorescence at 60°C. The gain of the laser will be a series of spikes, representing the allowed cavity modes, modulated with this profile. When the laser is operating, the spectral output will consist of a slice of the distribution, with a width dependent on the losses and the inversion density. For example, the output may consist only of the R_1 line if the gain for R_2, which is smaller, is insufficient to overcome the losses. Typically one might

† See footnote to Equation (2.24).

expect† an output pulse 0·2–0·3 nm wide ($\sim 10^3$ GHz) which will contain many cavity modes. A crystal, 10 cm long with integral reflectors, has a mode spacing of 840 MHz, and might produce about 10^3 cavity modes. Note, also, that the width of the spectral output will have a time dependence, reflecting the change in the inversion density as the pulse is first formed and then decays.

σ_{21} can be estimated from the fluorescence properties of ruby using Equation (5.11). For the R_1 line at 60°C, γ_t is approximately $1{\cdot}4 \times 10^3$ GHz, $\tau_{21}(R_1)(=1/A_{21})$ is about 3 ms, and thus $\sigma_{21}(f = 1) \simeq 2{\cdot}8 \times 10^{-20}\,\text{cm}^2$. According to Section 4.3, the rise time (for a perfectly switched system!) is determined by the parameters t_1 and t_2, which are inversely proportional to the cross section σ_{21}. t_2 is given by the equation

$$t_2 = \frac{V}{\sigma_{21}\, v \Delta N_0} \ln\left(\frac{\Delta N_0}{\Delta N_c}\right) = \frac{\eta}{\sigma_{21}\, c \Delta n_0} \ln\left(\frac{\Delta n_0}{\Delta n_c}\right) \tag{5.12}$$

The critical inversion density to sustain CW operation is, according to (4.52),

$$\Delta n_c = \frac{\Delta N_c}{V} = \frac{\delta_c}{\sigma_{21}\, 2L} \tag{5.13}$$

where, for simplicity, we put $L = L_c$.

Consider a laser consisting of a crystal 10 cm long $\times$ 0·3 cm^2, with integral reflectors, and assume that the only losses are from the output coupler. If the laser is operating on many modes, then it is realistic to take an average value of σ_{21} of around $2 \times 10^{-20}\,\text{cm}^2$, and use the single mode equations to estimate the pulse shape. For $\delta_c = 0{\cdot}1$, $\Gamma = 8{\cdot}5 \times 10^7\,\text{s}^{-1}$, if $\eta \simeq 1{\cdot}77$. The critical inversion, Δn_c, is 2×10^{17} per cm^3, and if the initial inversion is $n_0/2$, then $t_2 = 1{\cdot}3$ ns. According to (4.50) there will be a peak value of approximately 3×10^{17} photons in the cavity, and these will decrease by a factor 1/e in a time of $1/\Gamma$ or 11·9 ns. Of course the first build up will be determined by t_1, which will depend on the initial density of cavity photons, ϕ_0. In the complete on/off system this time will be much larger than t_2. It can be avoided if the change in Q is smaller, so that stimulated emission takes place during the pumping, as is the case in an active mode locked system.

Nd:YAG laser

This is one of the most widely used and versatile lasers. Although its output is in the infra-red part of the spectrum, it can produce much higher powers than an equivalent sized ruby. Basically, this is because it is a four level system and, as such, it is inherently more efficient. It is also more amenable to CW operation.

† The width of the output can be considerably reduced by cooling the crystal. This reduces the broadening γ_t.

The amplifying material is the colourless isotropic crystal $Y_2Al_5O_{12}$ (yttrium-aluminium garnet, YAG) in which about 1% of yttrium is replaced with Neodymium. The fluorescent properties are due to the energy levels of the Nd^{3+} ion. However, the level splitting in the crystal lattice is smaller than the ruby case, and it is convenient to classify the levels in the same way as the bare ion. In the rare-earths, the ground and excited states belong to configurations where the valence electrons occupy the $4f$ and $5d$ shells. Thus, the ground state configuration of Ce is $4f^1 5d^1$, and that of Nd, $4f^4 5d^0$. Divalent ions are formed by removing the two electrons from the closed $6s$ shell, and so they have essentially the same configurations, in the low lying states, as the atom. For trivalent ions, the third electron is removed either from the $5d$ or the $4f$ shell. The Nd^{3+} ion has a ground state configuration $4f^3$ and all the low-lying levels belong to this, or to the $4f^3 d^1$ configuration.

Figure 5.7 shows some of the relevant levels of Nd^{3+} in a YAG crystal. These will not, of course, be in the same positions in another host material, such as glass, or $CaWO_4$. The lowest levels from the $(4f)^3$ configuration have $S = 3/2$ and $L = 6$. According to the discussion in Section 1.6, the energy levels in LS coupling will be labelled $^4I_{9/2}$, $^4I_{11/2}$, $^4I_{13/2}$ and $^4I_{15/2}$. Above this, there are four levels, belonging to the ^4F term with $J = 3/2, 5/2, 7/2$ and $9/2$, and higher still are levels belonging to ^2H and ^2G terms. In a similar way to the ruby system, each of these

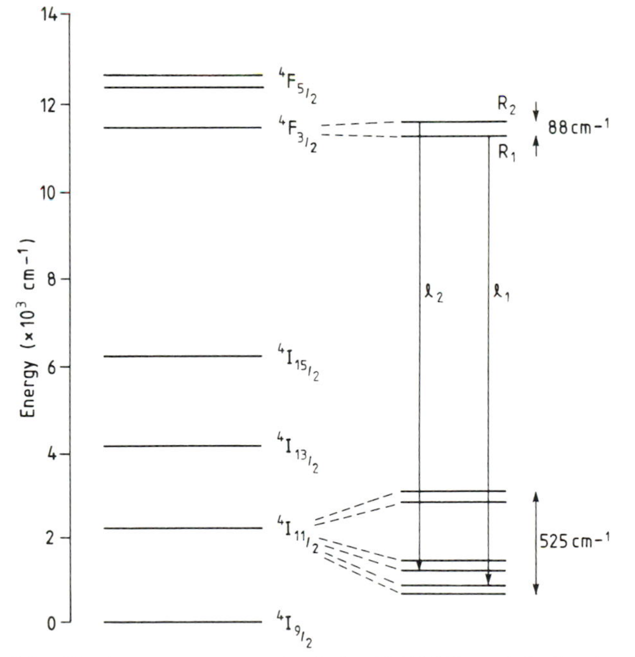

Figure 5.7. The important energy levels for Nd:YAG crystal, showing the origin of the two components, l_1 and l_2, of the 1·064 μm line.

levels is split into several components and homogeneously broadened because of the vibrations of the crystal lattice. Also, the selection rules for the decay of the bare ion no longer apply. Although the states have been roughly characterized by the total angular momentum, this quantity is no longer a good quantum number and transitions are possible, for example, between the $^4F_{3/2}$ and the $^4I_{9/2}$ levels. (This would normally involve an *E3* transition!) Single electron transitions of the type, $4f \rightarrow 4f$, are also possible.

The most important lasing transitions in the Nd:YAG system, are between the $^4F_{3/2}$ sublevels and the $^4I_{11/2}$ sublevels, as shown in Figure 5.7, giving a wavelength of 1·06 μm. Operation is also possible from the $^4F_{3/2}$ to the $^4I_{15/2}$ or $^4I_{13/2}$, but often, only if the dominant 1·06 μm emission is suppressed. (This can be achieved, in practice, using a frequency selective element in the cavity. Methods of cavity tuning are discussed later on.) A number of absorption bands above the $^4F_{3/2}$ level, can be used to provide the necessary population inversion, with the strongest around 0·75 and 0·81 μm (Figure 5.8). At room temperature, the $^4I_{11/2}$ is

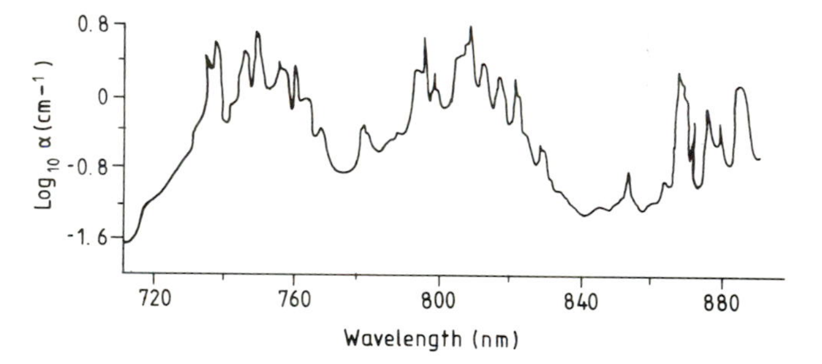

Figure 5.8. The absorption coefficient for Nd:YAG.
(Reprinted with permission from T. Kushida, H.M. Marcos and J.E. Geusic, *Phys. Rev.* **167** (1969) 289, © Bell Telephone Laboratories, 1968.)

largely unpopulated because the average lattice energy, kT, is much smaller than the energy of the lowest sublevel. In general, the pumping efficiency for Nd:YAG is less than ruby because the absorption bands are narrower. However, this is more than compensated for by the fact that σ_{21} is much larger and the inversion is easier to produce because the terminal laser transition is largely unoccupied until the light pulse is generated. Figure 5.9 shows the spontaneous emission spectrum from Nd:YAG at 1·06 μm. The broadening is asymmetric because it consists of more than one line, and, strictly, the problem is of an inhomogeneous nature. The equations will not, however, be the same as for Doppler broadening where a continuous variable is used to describe the different classes of atoms. Rather, in this case, the total gain will be a discrete sum (rather than an integral) from all the possible transitions. Also, a complete mathematical treatment would include the cross relaxation between the two states. In the ruby laser *R*1 and *R*2 are sufficiently far apart for them to be approximately considered as separate. The

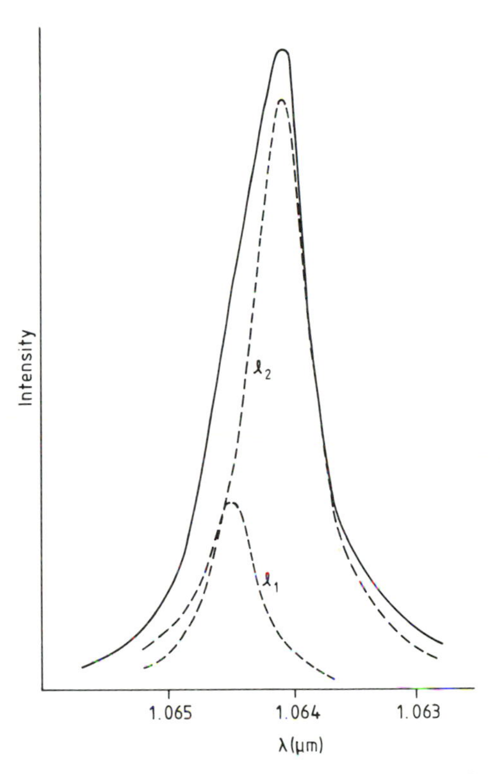

Figure 5.9. Emission spectrum of the 1·06 μm line in Nd:YAG at room temperature. This consists of two Lorenzian components, l_1 and l_2.
(From T. Kushida, H.M. Marcos and J.E. Geusic, *Phys. Rev.* **167** (1969) 289, © Bell Telephone Laboratories, 1968.)

total broadening of Nd:YAG at room temperature, is similar to the ruby fluorescence, so the increase in the value of σ_{21} must be due to the shorter fluorescent lifetime (see Equation (5.11)). Since $\sigma_{21} \sim 6\times10^{-19}$ cm this lifetime must be of the order of fractions of a ms. For a pulsed system, increasing σ_{21} is one way to improve the gain, but this cannot go on indefinitely, because at some point, the rate of spontaneous emission (n_2/τ_{21}) will start to compete with the pumping rate preventing the formation of a population inversion. For efficient CW operation it is important that the total lifetime for return to the ground state, τ_2, should be shorter than the lifetime of the laser transition, τ_{32}. Indeed, for weak pumping, Equation (4.27) shows that this condition is imperative, if $g_1 = g_2$. For Nd:YAG all of the ^4I levels are in thermal equilibrium with the crystal lattice, and transitions from the largely unoccupied $^4I_{11/2}$ to the ground state, take place by fast phonon relaxation, with lifetimes, τ_2, less than 10^{-6} s. Similarly, the pump levels are in thermal equilibrium with the $^4F_{3/2}$ levels. The $^4F_{3/2}$ to ^4I transitions, which would require several phonons, have long lifetimes ($\tau_{32} = \tau_{1\cdot06} = 550\,\mu$s) and decay by photon emission.

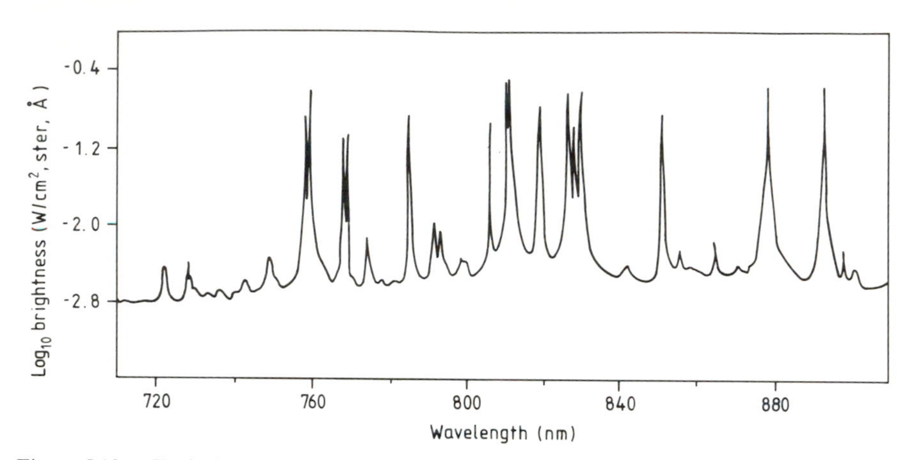

Figure 5.10. Typical output intensity, as a function of wavelength, of a krypton arc lamp. (From W. Koechner, L.C. DeBenedictis, E. Matovich and G.E. Meyer, *IEEE J. Quant. Electron.*, QE-**8** (1972) 310, © IEEE, 1972.)

An efficient method for optical pumping of Nd:YAG crystals in CW systems is using a krypton filled arc lamp. The strongest output lines from this type of source conveniently overlap the absorption bands, as Figure 5.10 and 5.8 show. Figure 5.11 shows a simple scheme for a mode locked laser. Cavity modulation is effected using an acousto-optic modulator, though this could equally be an electro-optic device. When high power pulsed operation is required the system would be similar except that a flash lamp would replace the arc lamp, and some

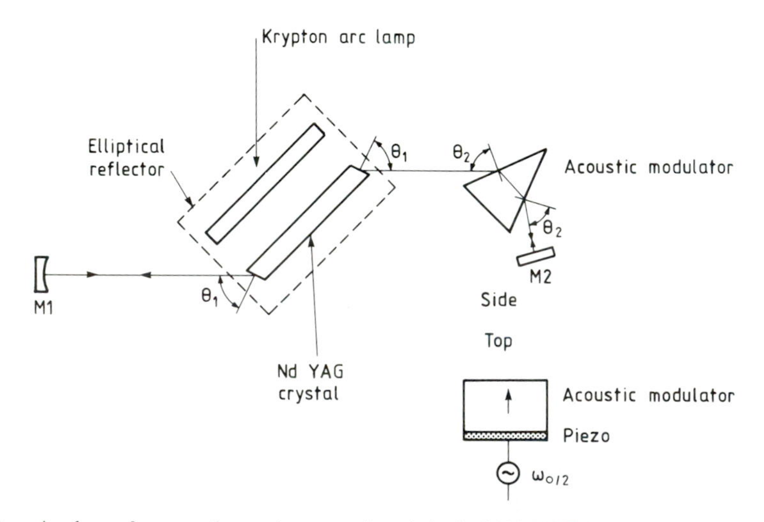

Figure 5.11 A scheme for a continuously pumped mode locked Nd:YAG laser using an acoustic-optic modulator. θ_1 and θ_2 are Brewster angles for the Nd:YAG crystal and the modulator. This arrangement ensures no reflection losses for linearly polarized light.

type of Q-switching would be required. Another way to increase the output power of any laser (for a short time) is by cavity dumping. In this process the laser beam in the cavity is suddenly extracted using some form of optical deflector, which is switched from its normal position of allowing the beam to travel back and forth between the mirrors to deflecting the beam out of the cavity. The concept is very similar to the Q-switching idea, except that now the Q is suddenly reduced. In an ideal system, it would amount to instantaneously adjusting the output coupler from a very low transmission to very high one, thus allowing all the cavity photons to escape in an approximate time of $2L_c\eta/c$. The dumping can be understood in terms of the analysis of Section 4.3 for a Q-switched system. For an ideal system with no losses the number of cavity photons, after the switch, will approach the value $\Delta N_0/2$. Now the Q is switched to a low value by means of a cavity dumper. In the previous case the pulse decay length was

$$\frac{1}{\Gamma} = \frac{2L_c\eta}{\delta_c c} \tag{5.14}$$

when only a fraction δ_c of the light was extracted in the time it takes the light to make a complete round trip. Now δ_c is effectively one so this time is much shorter. A further advantage is that the absorption and scattering losses are in principle smaller if the device itself has low losses. This is because the total distance travelled by the laser beam is reduced by a factor $1/\delta_c$.

Because δ_c is large in most flashlamp pumped systems†, there is little to be gained from using a cavity dumper particularly when both the response time and losses of the system are considered. In practice, cavity dumping is most often employed in continuously pumped systems with high Q values (low losses). For example, the mode locked Nd-YAG system shown in Figure 5.11 can be switched so that the single circulating pulse is extracted from the cavity. If the switching can be made instantaneous, then the peak intensity of the pulse is increased by a factor $1/\delta_c$ from a system with matched output losses of δ_c. Similar systems can be employed with CW gas lasers. Beam deflection in the cavity can be achieved with electro-optic or acousto-optic deflectors. These devices (as do the modulators and Q switches mentioned earlier) depend for their operation on changes in the refractive index of a material induced by either an electric field or an acoustic wave‡. Since these play an important role in laser systems, they are briefly described here, although more detailed descriptions of the various types can be found elsewhere.

5.2. Optical modulators

In the acoustic modulator (or deflector) an acoustic wave is generated in an optically transparent material, such as a crystal of tellurium dioxide, using an

† If the gain of the system is high the output coupling can be large.

‡ Modulators can also be made using the Faraday effect (e.g. YIG modulator).

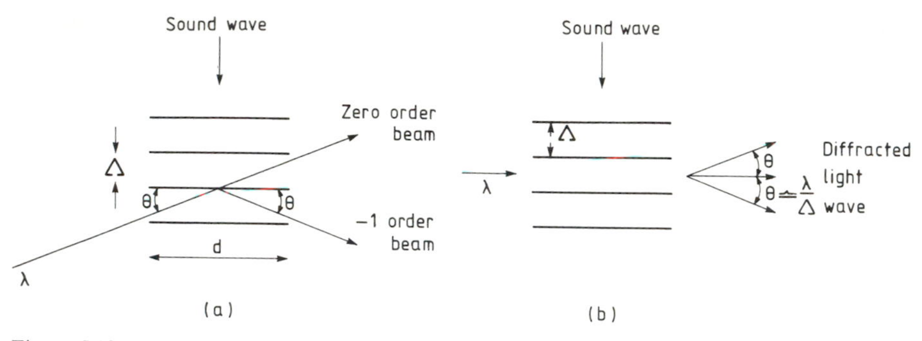

Figure 5.12. Two types of acousto-optic modulator.

ultrasonic transducer. This wave gives rise to periodic changes in density which give rise to regular alterations of the refractive index. In the Bragg angle modulator (or deflector), shown in Figure 5.12(a), a set of parallel wavefronts is induced normal to the direction of sound propagation, and the light beam is incident at the Bragg angle

$$\sin\theta = \frac{\lambda}{2\eta\Lambda} \tag{5.15}$$

where Λ is the wavelength of the compressional wave. The light beam is reflected at an angle θ from the wavefronts. This is similar to the reflection at the interface between two different dielectrics, except that the variation in refractive index is not discontinuous, but a smooth function. The intensity of the 'scattered' wave will depend on the number of intercepts of the light beam and the acoustic wavefronts, because the Bragg criterion ensures that successive reflections will be in phase. This number is just

$$N = d\frac{\tan\theta}{\Lambda} \tag{5.16}$$

where d is the length of the crystal. Since $\lambda/\Lambda \ll 1$, then $\sin\theta = \tan\theta = \theta$, and Equation (5.15) can be substituted into (5.16) to give

$$N \simeq \frac{\lambda d}{2\eta\Lambda^2} = \frac{\lambda d f^2}{2\eta u_0^2} \tag{5.17}$$

where f is the acoustic frequency and u_0 is the velocity of the sound wave. Typically, u_0 might be about $1{\cdot}5 \times 10^5$ cm/s, so for visible light, and dimensions of the order of 10 cm, f must be greater than 10 MHz. Now the fraction of light scattered from each wavefront will depend on the amplitude of the sound wave, since this will determine the variations in density, which can be related to refractive index changes. For many crystals this fraction is fairly large, at the Bragg angle, so that a single intercept is sufficient unless complete modulation is required.

The reflected light wave is also Doppler shifted in frequency since it scatters from a moving medium. The fractional increase in frequency is

$$\frac{\Delta\nu}{\nu} = \pm 2u_0\eta\sin\theta/c \tag{5.18}$$

where the sign depends on the relative directions of the sound and light wave. A factor of 2 is included because, in the reference frame of the travelling wavefront, the incident beam is already shifted by one half of this amount. Using (5.15), this becomes

$$\Delta\nu = \pm f \tag{5.19}$$

where, again, f is the acoustic frequency.

The acoustic modulator used to mode lock the Nd:YAG system, of Figure 5.12, is somewhat different in concept. Here, the travelling acoustic wave is used like a normal-incidence diffraction grating (Figure 5.12(b)), with the incident light parallel to the wavefronts. The two first-order diffracted beams will be at angles, given by the equation

$$\sin\theta = \pm\lambda/\Lambda \tag{5.20}$$

and the Doppler shift in frequency is

$$\Delta\nu = \pm\nu u_0\sin\theta/c = \pm f \tag{5.21}$$

In the acoustic mode locker, the acoustic frequency is set at exactly one half the cavity mode spacing. When the beam passes through the modulator, a certain fraction of it (which depends on the intensities of the first-order diffracted beam) is shifted down in frequency by half the mode spacing, and the same fraction is shifted upwards in frequency by the same amount. After reflection from the mirrors these shifted components are either returned to the original mode frequency, or further shifted by half the mode spacing. A round trip of any single mode in the cavity gives rise to sidebands at exactly the positions of the other modes, and this effectively locks the modes together with a common phase (see Section 4.4). Since the modulator introduces a small angular spread in the beam, the optics are most favourable when it is placed at the beam waist.

Electro-optic modulators depend on the changes in birefringence introduced in certain materials under the infleunce of an applied electric field. When the refractive index changes are linear with applied field, it is known as the Pockels effect, whilst a quadratic variation is known as the Kerr effect. Physically,

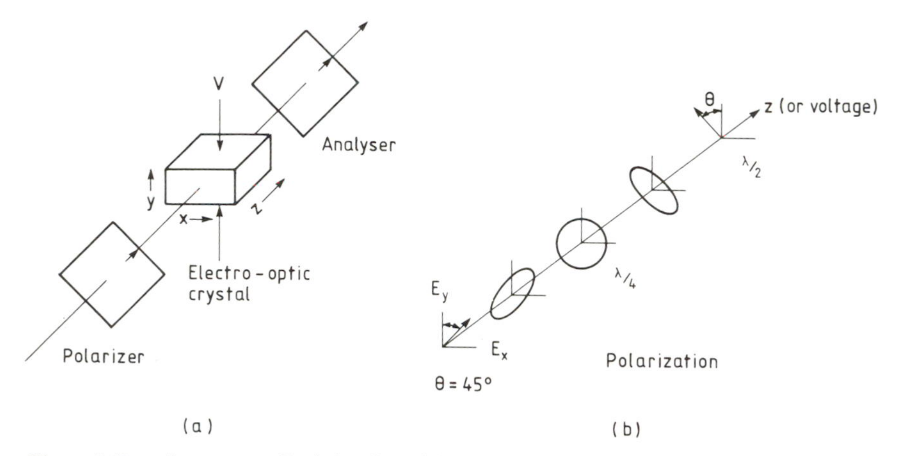

Figure 5.13. A transverse Pockels cell modulator (a), and the changes in polarization (b) as the light passes through the crystal.

this alteration is quite simple to understand though a general mathematical treatment is rather involved since it requires the use of tensor quantities. A simple picture is that there is a change in the static polarization along the direction of the applied field, which can be envisaged as an alteration in the microscopic charge distributions. This affects the dynamic polarization, which, in turn, is related to the refractive index.

A common type of device is the transverse Pockels cell, shown in Figure 5.13. The electro-optic material is often a crystal (for example, lithium tantalate, $LiTaO_3$) which is cut along certain planes so that an electric field applied along one axis introduces birefringence for light propagating along an orthogonal axis. Consider the geometry shown in the figure, with the laser beam travelling along the z direction and linearly polarized at 45° to the x and y axis by the presence of the first polarizer. In the absence of any birefringence the beam is unaffected and passes through the second polarizer. When an electric field is applied, the two components of the radiation, E_x and E_y, are affected differently. In essence, the wavelength components, λ_x and λ_y, now become λ/η_x and λ/η_y, corresponding to the two different phase velocities. After traversing a certain length, l, the state of polarization is dependent on the total phase shift between the two components. If the difference, or retardation

$$\Delta\lambda = \lambda l(1/\lambda_x - 1/\lambda_y) = l(\eta_x - \eta_y) \tag{5.22}$$

is $\pm\lambda/2$ (corresponding to one extra half cycle of the field in the x or y planes) the emerging wave is still linearly polarized but the direction of polarization is rotated through an angle of $\pi/2$ in the xy plane. On the other hand, for $\Delta\lambda = \pm\lambda/4$, the output is circularly polarized. In general, the polarization goes through the sequence shown in the figure. The important characteristic of the material can be defined by the half wave voltage,

$$v_{\lambda/2} = (El)_{\lambda/2} \tag{5.23}$$

where $(El)_{\lambda/2}$ is the product of the field and length required to give a shift of $\lambda/2$. The voltage, required to give a $\lambda/2$ shift, on a modulator of length, L, and width, W, is therefore

$$V_{\lambda/2} = v_{\lambda/2}\frac{W}{L} \tag{5.24}$$

Typically, a few thousand volts would be required on a crystal when $W = L$ ($v_{\lambda/2} = 2{\cdot}8$ kV at $\lambda = 630$ nm, for $LiTaO_3$).

The operation of the modulator is now straightforward. For a Q-switch, with a complete on/off characteristic, the voltage is switched by an amount given by (5.24). With the polarizers as shown in the figure, the crystal is held at voltage during the pumping, and when the pumping is completed the voltage is reduced as rapidly as possible. Amplitude modulation can be achieved by applying an alternating voltage across the crystal. With suitable circuits it is possible to modulate at frequencies up to 1 GHz, this being determined by the speed at which the voltage can be applied rather than the response of the electro-optic crystal. The modulated intensity (for a Pockels cell) is given by the equation

$$I = I_0 \cos^2\left(\frac{\pi V}{2V_{\lambda/2}}\right) \tag{5.25}$$

where I_0 is the maximum intensity and V is the applied voltage.

5.3. *Semiconductor lasers*

Semiconductor lasers are quite different in concept to the previous solid lasers, because the energy levels cannot be said to belong to any atom, ion or molecule. Together with free electron lasers, they are unique in this characteristic. The highest energy levels are a property of the whole crystal lattice, and the electrons which occupy them are not localized with respect to any one atom. Of course, the more tightly bound electrons, belonging to the inner shells, are relatively unaffected by the crystal structure. The general features of the energy levels for an intrinsic semiconductor are shown in Figure 5.14. These show two broad band structures separated by an energy gap. Below this gap there is the valence band and above it the conduction band, with the probability of an electron being in a given energy level, described by the Fermi-Dirac function

$$f_E = \frac{1}{1+\exp((E-E_f)/kT)} \tag{5.26}$$

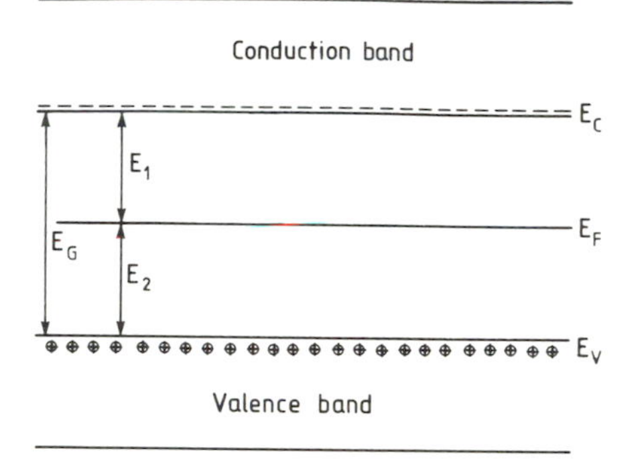

Figure 5.14. Band structures in a semiconductor.

Here f_E is the probability and E_f is the Fermi energy, or chemical potential.

Now the distinguishing features of a semiconductor, as opposed to a metal or insulator, lie in the relative occupancy of these general band structures. For a semiconductor, at absolute zero, the conduction band is completely empty and the valence band is full, whilst at elevated temperatures a certain number of electrons will be in the conduction band. Promotion of electrons into the conduction band leaves energy levels unoccupied in the valence band, thus giving rise to holes or vacancies. These holes have a charge equal in magnitude, but opposite in sign, to the electron, and in general, their mass is different to the free electron. (In the general analysis, both the electrons in the conduction band, and the holes in the valence band are described by an effective mass parameter. These masses determine, for example, the mobilities in an electric field, and they are not necessarily the same for electrons and holes.)

The important quantity to calculate, for laser theory, is the number density of electrons (and holes) as a function of energy. This is simply the product of the occupation probability (5.26) and the density of states. The latter can be calculated in an analogous way to the photon density of states given in Section 2.1. However, for electrons in solids, the boundary conditions, on the allowed momenta, are a real property of the crystal lattice rather than a mathematical construction. The allowed values of the wavevector are

$$k_x = \frac{n_x \pi}{L_x} \qquad k_y = \frac{n_y \pi}{L_y} \qquad k_z = \frac{n_z \pi}{L_z} \tag{5.27}$$

and the density of states, calculated in the same manner as previously is

$$\rho(k)\, dk = \left(\frac{k^2}{\pi^2}\right) dk \tag{5.28}$$

Here the factor of two now arises from the two spin states of the electron rather than the two states of polarization. Equation (5.28) can be transformed into energy space by noting that for electrons in the conduction band†

$$E = E_c + \frac{\hbar^2 k^2}{2m_e} \tag{5.29}$$

where E_c is the energy of the bottom of the band, and m_e is the effective electron mass. Similarly, the equation relating the hole energy and wavenumber is

$$E = E_v - \frac{\hbar^2 k^2}{2m_h} \tag{5.30}$$

where, again, E_v is the energy at the top of the valence band, and m_h is the effective hole mass. The density of states for electrons and holes is, therefore,

$$\rho_{eh}(E)\,dE = \frac{1}{\pi^2}\left(\frac{m_{eh}}{\hbar^2}\right)^{3/2} (2|E - E_{cv}|)^{1/2}\,dE \tag{5.31}$$

$\rho_{eh}(E)\,dE$ is then the total number in the interval from E to $E + dE$. The electron number density in the conduction band is, therefore, (5.26) multiplied by ρ_e, above. For holes the situation is slightly different because we need to multiply the density of states by the probability of their being a hole in the valence band. This is not f_E, but $1 - f_E$, or

$$f_{Eh} = \frac{1}{1 + \exp((E_f - E)/kT)} \tag{5.32}$$

So, finally, the number of densities of electrons and holes are given by the equations

$$n_e^c(E)\,dE = \frac{1}{\pi^2}\left(\frac{m_e}{\hbar^2}\right)^{3/2} (2|E - E_c|)^{1/2} \frac{1}{1 + \exp((E - E_f)/kT)}\,dE \tag{5.33}$$

and

$$n_h^v(E)\,dE = \frac{1}{\pi^2}\left(\frac{m_h}{\hbar^2}\right)^{3/2} (2|E - E_v|)^{1/2} \frac{1}{1 + \exp((E_f - E)/kT)}\,dE \tag{5.34}$$

These two equations give the number of electrons (and holes) per unit volume, in the energy interval from E to $E + dE$. They can be integrated with respect to energy to give the total numbers of electrons and holes in the conduction and

† This quadratic approximation applies to states near the value E_c, or E_v for holes.

valence bands respectively. Since these must be equal, the Fermi energy can only be midway between E_c and E_v, for the unrealistic conditions $m_h = m_e$. However, if $E_g \gg kT$, the exponential part of the Fermi-Dirac function is dominant, and the Fermi energy always lies within about kT of the centre of the band gap. At room temperature (300 K), kT is 0·026 eV, which can be compared with band gap energies of 1·12 eV for silicon, 0·75 for germanium, and 1·4 eV for gallium arsenide.

For an intrinsic semiconductor in thermal equilibrium, where $kT \ll E_g$, the electron and hole densities, (5.33) and (5.34), can be approximated to

$$n_e^c(E) \simeq \frac{\sqrt{2}}{\pi^2}\left(\frac{m_e}{\hbar^2}\right)^{3/2} \exp\left(-\frac{E_1}{kT}\right)(\Delta E)^{1/2} \exp\left(-\frac{\Delta E}{kT}\right) \tag{5.35}$$

and

$$n_h^v(E) \simeq \frac{\sqrt{2}}{\pi^2}\left(\frac{m_h}{\hbar^2}\right)^{3/2} \exp\left(-\frac{E_2}{kT}\right)(\Delta E)^{1/2} \exp\left(-\frac{\Delta E}{kT}\right) \tag{5.36}$$

where E_1 and E_2 are shown in Figure 5.14, and $\Delta E = E - E_c$, for electrons, or $\Delta E = E_v - E$ for holes. The hole distribution is therefore the mirror image of the electron distribution (Figure 5.15) and, in equilibrium,

$$(m_e)^{3/2} \exp\left(-\frac{E_1}{kT}\right) = (m_h)^{3/2} \exp\left(-\frac{E_2}{kT}\right) \tag{5.37}$$

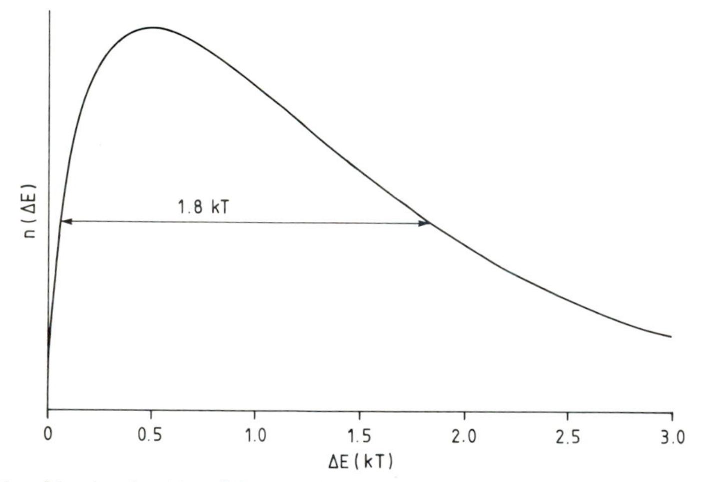

Figure 5.15. Number densities of electrons (holes), in the conduction band (valence band) per unit energy interval as a function of the energy, $\Delta E = E - E_c$ ($\Delta E = E_v - E$).

The number density of electrons in the valence band is then approximately

$$n_e^v(E) = \frac{\sqrt{2}}{\pi^2}\left(\frac{m_e}{\hbar^2}\right)|E - E_v|^{1/2} \tag{5.38}$$

which exceeds the number of holes by the factor, $\exp(E_1 + \Delta E)/kT)$, or, approximately, $\exp(E_g/kT)$.

At any temperature electrons in the conduction band can fall back into vacant sites, or holes, in the valence band. This decay can be accompanied by the emission of photons (pure electromagnetic), by a combination of photons and phonons or by phonons alone. The relative amounts of each process will depend on the energy gap, the temperature and the type of semiconductor. Two classes of semiconductor can be distinguished. In direct semiconductors the minimum value of energy (in an energy/momentum distribution), in the conduction band, occurs at the same momentum as the maximum in the valence band. On the other hand, for indirect semiconductors the two are shifted. Since we are considering transitions which take place close to E_g, those involving indirect semiconductors require larger changes in momentum, typically, $h/2a$, where a is the crystal lattice dimension (assuming a simple one-dimensional model). This cannot be supplied by a photon whose momentum is h/λ_g $(\lambda_g/a \gg 1)$. Here the transition takes place by the emission of a photon and a phonon, such that the energy change is

$$\Delta E = \hbar\omega_p + \hbar\omega \tag{5.39}$$

where $\hbar\omega_p$ is the phonon energy (which is much smaller than the photon energy $\hbar\omega$). Because the decay is a two step process it is much slower than the direct one, and hence the cross section is smaller.

Absorption of photons can take place by the reverse process of transferring an electron from the valence band to the conduction band. This absorption will have a broad energy distribution because the electron can be promoted from any point in the valence band to any unoccupied level in the conduction band. The electron can then lose energy by phonon emission until it falls to the 'top' of the occupied levels in the band. These interband relaxation processes are extremely rapid and more than compete with recombination processes. Similarly, the vacancy or hole in the valence band will rapidly be in thermal equilibrium with the other holes in the band. An optically pumped laser could then be envisaged where the pumping depletes the top of the valence band whilst populating the bottom of the conduction band, with the recombination radiation providing the laser transition. Such a scheme might be feasible if the temperature and band gap were such that the thermally occupied levels were close to E_c for electrons and E_v for holes. (This is when (5.35) and (5.36) are valid.) Recombination radiation would then be centred in a narrow line with an energy slightly larger than E_g. However, this is essentially a three level system and, in fact, it is more advantageous to use a doped semiconductor (p type) which is akin to a four level operation. This is described later on. In fact, the bulk absorption losses in all semiconductor lasers are much larger than in doped insulators so that only very high gain schemes will operate. This absorption is due to electron (or hole transitions) within each band, and is known as free carrier absorption. Ultimately, this energy is transferred to heat because the decay within the band, when the electron returns to thermal equilibrium, is by rapid phonon emission.

The criterion for light amplification in semiconductors is just the same as before; namely that the stimulated emission should exceed the absorption. In the analysis of other types of lasers the number densities of atoms in thermal equilibrium was represented by Maxwell-Boltzmann statistics, because there was no restriction on the numbers in any particular energy level. Now we have a different situation where each energy level can accommodate only two electrons at most. However, the levels are so closely packed that they form wide band structures. In any electromagnetic transition we must consider all the possible different energy routes for both emission and absorption. It is this energy spread which is the major cause of broadening in semiconductors. Within each band the levels are in thermal equilibrium and the broadening can be considered homogeneous in nature. This can be thought of as very rapid cross relaxation, where any state depleted by stimulated (or spontaneous) emission is filled so rapidly that hole burning is not possible.

Before calculating the necessary conditions required for amplification, we first need to calculate the number of densities of electrons when the conduction and valence bands are not in thermal equilibrium. (The two equations (5.33) and (5.34) are, of course, for thermal equilibrium.) In general, the number density of electrons in the conduction band, and holes in the valence band, are

$$n_e^c = \frac{1}{\pi^2}\left(\frac{m_e}{\hbar^2}\right)^{3/2}(2|E-E_c|)^{1/2}\frac{1}{1+\exp((E-E_{fc})/kT)} \tag{5.40}$$

and

$$n_h^v = \frac{1}{\pi^2}\left(\frac{m_h}{\hbar^2}\right)^{3/2}(2|E-E_v|)^{1/2}\frac{1}{1+\exp((E_{fv}-E)/kT)} \tag{5.41}$$

where E_{fc} is the 'quasi Fermi' energy for the conduction band and E_{fv} is that for the valence band. Also, the number of densities of holes in the conduction band and electrons in the valence band can be simply obtained by changing the sign of the exponential term. The fact that the two functions can be written in the same form as previously implies that each band is separately in thermal equilibrium during the inversion. This means that the phonon processes within each band are much more rapid than the band to band processes.

How can this non thermal distribution be produced, and what are the conditions on E_{fc} and E_{fv} for light amplification? One way, which has already been mentioned, is by optical pumping. (Electron beams can also be used to transfer electrons from the valence to the conduction band.) Initially, with no pumping the quasi levels will be the same as E_f. Under pumping the levels diverge, and E_{fc} moves to the conduction band whilst E_{fv} decreases to the valence band. If the pumping is strong enough the quasi levels will penetrate the bands. At this point it is possible for laser action to start, as the later analysis will show. Another way is using a forward biased *p–n* junction. In any semiconductor it is possible to introduce dopant atoms (impurities) which radically alter the electron distribution in the bands. The effects depend on whether the dopants remove electrons or add them. For *n* type semiconductors, the added elements introduce more electrons than the lattice site calls for. The Fermi energy moves upwards towards the conduction band as the donor impurity level is increased, and will eventually penetrate the band at high enough concentrations. When this happens the semiconductor is said to be degenerate. In *p* type material the added impurities remove electrons so the Fermi energy is reduced. If a *p–n* junction is formed, the energy bands at the junction become displaced so as to maintain the same Fermi energy on both sides of the junction, and hence maintain thermal equilibrium (Figure 5.16(a)). Applying a forward bias (Figure 5.16(b)) displaces the 'Fermi' energy levels by an amount equal to the applied voltage (less any ohmic losses if a current flows). If this voltage is increased then a point is reached where electrons in the conduction band of the *n* type material can be injected into the conduction band of the *p* type material, and holes in the *p* type material can be injected into the valence bands of the *n* type semiconductor. As soon as this happens recombination radiation is produced in a small region, depth *d*, at the interface, and a current flows in the circuit. If the current in the system is increased a point can be reached where, for suitable geometries, light amplification is possible. This is the basic principle of an injection laser. For this non-thermal equilibrium, equations (5.40) and (5.41) can be used for the number density of electrons and holes.

The basic condition for amplification is that the absorption of light should

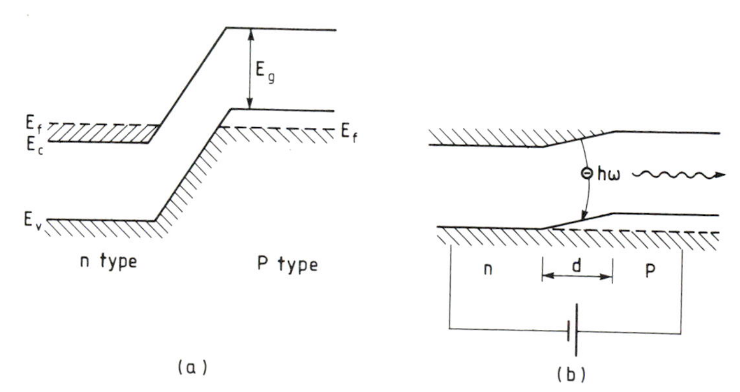

Figure 5.16. A *p–n* junction (a), and the effect of a forward bias (b).

be smaller than the stimulated emission. This puts a constraint on the values of the quasi Fermi energies, as the following analysis shows. The stimulated emission rate (number of quanta emitted per unit time per unit volume) from the levels at E_e, in the conduction band, to those at E_n, in the valence band, will be

$$SE = CB_{cv} f_c(E_e)(1 - f_v(E_h))W(E_e - E_h) \tag{5.42}$$

Here W is the photon energy density at frequency $\omega = (E_e - E_h)/\hbar$. B_{cv} is the conduction band to valence band rate coefficient, and C is a constant which incorporates a product of the initial and final density of states. Similarly the rate of absorption from E_h to E_c is

$$AB = CB_{vc} f_v(E_h)(1 - f_c(E_e))W(E_e - E_h) \tag{5.43}$$

The criterion for amplification (when there are no losses!) is simply that

$$SE > AB \tag{5.44}$$

Since $B_{cv} = B_{vc}$ this inequality can be written in terms of the Fermi-Dirac functions for electrons in the two bands. Thus

$$f_c(E_e)(1 - f_v(E_h)) > f_v(E_e)(1 - f_c(E_h)) \tag{5.45}$$

where, again, c and v refer to the conduction and valence bands. Equation (5.45) is reduced to

$$f_c(E_e) > f_v(E_h) \tag{5.46}$$

This is equivalent to the population inversion requirement derived in Section 3.1. Using (5.26) it can be written

$$\exp(E_e - E_{fc}) < \exp(E_h - E_{fc}) \tag{5.47}$$

or

$$E_e - E_h < E_{fc} - E_{fv} \tag{5.48}$$

and, therefore

$$E_{fc} - E_{fv} > E_g \tag{5.49}$$

The levels below E_{fc}, in the conduction band, are largely occupied, whilst those above E_{fv} in the valence band are largely empty. With reference to Figure 5.16, amplification is only possible for electron transitions from the 'shaded' region of the conduction band to the 'empty' region of the valence band.

It is instructive to calculate the gain of a semiconductor laser in a rigorous manner. To do this we use the general QED expression (2.193)[†], which can be written, for a conduction band to a valence band transition, as

$$P_{cv} = P_0\,\delta(\hbar\omega - E_e + E_h)\cdot\begin{matrix}\phi \\ \phi+1\end{matrix} \tag{5.50}$$

Here, ϕ, which represents the total density of photons, is used for absorption, and $\phi+1$, for spontaneous or stimulated emission. In writing P_0 as a constant, we are assuming that the matrix elements are the same for any of the two states, and that ω does not vary very much over the gain profile. The most convenient expression for the gain when using this expression is

$$\alpha = \frac{1}{I}\frac{dI}{dz} = \frac{1}{\phi}\frac{d\phi}{dt}\frac{dt}{dz} = \frac{\eta}{c}\frac{1}{\phi}\frac{d\phi}{dt} \tag{5.51}$$

The gain is calculated by summing over all the states in the conduction and valence bands allowing for the occupation probabilities as determined by the Fermi-Dirac distributions. Thus the total rate of change of photon density (omitting the implicit dependence on E_e and E_h) is

$$\frac{d\phi}{dt} = \int P_{cv}[\rho_c f_c(1-f_v)\rho_v - \rho_v f_v(1-f_c)\rho_c]\, dE_e\, dE_h \tag{5.52}$$

where the first term, in the square brackets, is the stimulated emission and the second one is absorption. (If the calculation had been for atoms or molecules (bosons!) then the two terms would just be the number densities of atoms in the upper and lower levels.) Substituting (5.50) into (5.52), and cancelling cross terms, $f_c f_v$, gives

$$\alpha(E) = \frac{\eta}{c}P_0 \int \rho_c(E)\rho_v(E-\hbar\omega)[f_c(E) - f_v(E-\hbar\omega)]\, dE \tag{5.53}$$

This expression looks in the first instance, to be totally different to the familar expression (3.11) for bosons. However, this is only because the number densities and homogeneous broadening are rolled together inside the integral. To write the gain in a more recognizable form, the constant P_0 can be related to the average electron lifetime for recombination. Again, the expression (5.50) is used but, for spontaneous emission, only one photon is involved, and we need to sum

† Note that P_{cv} is a probability per unit time, or a rate.

over all the final hole states, as well as including the photon density of states†. The A coefficient, or inverse lifetime, is therefore

$$\frac{1}{\tau} = \rho(\hbar\omega)P_0 \int \rho_v(E_h)(1 - f_v(E_h))dE_h = \rho(\hbar\omega)n_h P_0 \tag{5.54}$$

where $\rho(\hbar\omega)$ is the energy density of photon states at the transition frequency ($\omega = \omega_{cv} \simeq E_g/\hbar$), and n_h is the total hole density in the valence band. Furthermore, it is possible to approximate the integral in (5.53) when the semiconductor is cooled. In these circumstances the electrons are injected into a narrow energy band which is assumed to be small compared with the hole distribution. The gain for transitions where f_v is zero in (5.53) is then

$$\alpha(\hbar\omega) \simeq \left(\frac{\eta}{c}\right) P_0 \, n_e \, \rho_v(E_0 - \hbar\omega) \tag{5.55}$$

where E_0 is the energy of the electrons in the conduction band ($\simeq E_c$) and n_e is the number density of electrons in the conduction band. If the gain profile is roughly triangular in shape then

$$n_h \simeq \rho_v(\max)\Delta E \tag{5.56}$$

where ΔE is the full width (at half height) of the homogeneous broadening. Combining (5.56), (5.55) and (5.54) gives

$$\alpha_{\max} \simeq \frac{\eta n_e}{c\tau\rho(\hbar\omega_0\Delta E} \tag{5.57}$$

for the peak gain at low temperatures.

Evidently, the 'inversion' is just n_e, and using the equation

$$\rho(\hbar\omega_0) = \frac{\eta^3\omega^2}{\pi^2\hbar c^3} \tag{5.58}$$

for the photon density of states, the cross section is reduced to

$$\sigma_{\max} \simeq \frac{\pi^2 c^2}{2\tau\eta^2\omega^2\gamma_t} \tag{5.59}$$

where γ_t is the half width of the gain profile in frequency. Apart from the small difference in numerical factors this is identical to that derived for a homogeneously broadened atomic system (Equation (5.11)).

† See the analysis in 2.10.

It is easy now to see why the gain is so much larger in a direct semiconductor like GaAs (the common *p–n* junction material) than in a doped insulator. In both, the homogeneous broadening is of the same order, but the lifetime in the semiconductor is 10^{-9} s, or approximately 10^6 times that in ruby. This means that the cross section in a semiconductor is much larger although this is partly offset by a smaller 'inversion' density. However, as we mentioned earlier, the losses in a *p–n* junction are much higher. The line broadening in a semiconductor is determined by the extent of the doping and the temperature. Since it is necessary for the semiconductor to be degenerate, the width of the hole distribution, and consequently that of the gain profile, must be at least of the order kT (see Equation (5.32)). Cooling the semiconductor is beneficial in increasing the gain, but, because the currents in a *p–n* junction are large, this cannot be carried too far. Figure 5.17 shows the output characteristics of a GaAs injection laser.

One of the major drawbacks in a single junction laser is the comparatively narrow region in which the inversion can be maintained. The width, d, (Figure 5.16) is essentially determined by the distance the electrons or holes diffuse before recombination. If D is the diffusion coefficient then

$$d = (D\tau)^{1/2} \tag{5.60}$$

For the case of a heavily doped GaAs junction, d might be typically 2 μm for electrons. (D for holes is much smaller and can usually be ignored.) It is now obvious why the diffraction losses are so large. The minimum angular spread (see for example, the diffraction of a plane wave passing through a slit), in any system with dimension, d, is

$$\theta \sim \lambda/d \tag{5.61}$$

and since, for GaAs, $\lambda = 0{\cdot}84\,\mu$m, the diffraction losses are extremely large. This fact, coupled with the high free carrier absorption, means that the Q values are low. There is, therefore, little point in making a very long cavity or having high reflectance mirrors. In fact, the usual configuration is to polish the end faces of the GaAs crystal since the high refractive index allows about 35% reflection back into the cavity. A typical geometry might consist of a crystal, in the form of a regular parallelepiped, with the junction having an area of 1 mm × 0·1 mm. The crystal is polished so that the stimulated emission is parallel to the longest side. Although one might expect the homogeneous nature of the system to restrict the oscillation to a single mode this is not the case. This is because the spatial hole burning plays an important role in mode competition.

Equation (5.57) for a cooled GaAs junction can be more conveniently written in terms of the current in the diode, since

$$n_e = \left(\frac{\beta J}{ed}\right)\tau \tag{5.62}$$

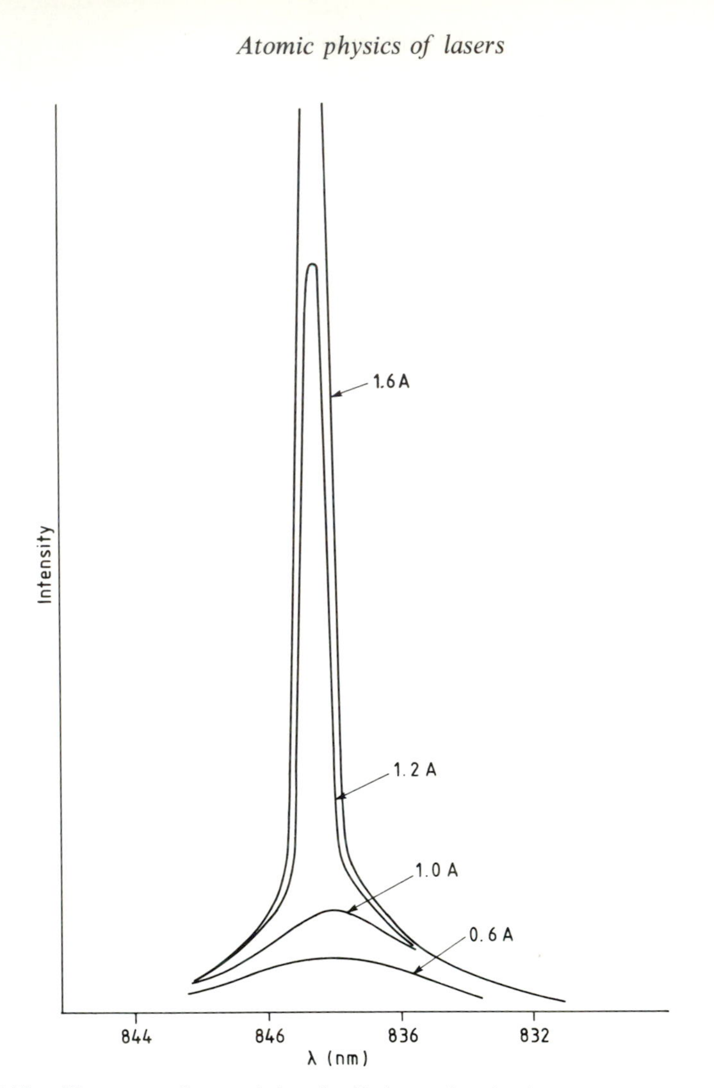

Figure 5.17. The output characteristics of a GaAs *p–n* junction laser.
(From G. Burns and M.I. Nathan, *Proc. IEEE* **51** (1963) 471, © IEEE, 1963.)

Here β is the quantum efficiency and J is the electron current density. Substituting (5.62) into (5.57), and using (5.58) gives

$$\frac{\alpha_{\max}}{J} = \frac{\pi^2 \beta^2 c^2}{2ed\eta^2 \omega^2 \gamma_t} \tag{5.63}$$

for the gain per unit current density. For typical values of $d = 2\,\mu\text{m}$, $\beta = 0{\cdot}7$ and $\gamma_t = (2\pi)2{\cdot}4\,\text{GHz}$, this gives $\alpha/J = 2\,\text{mm/A}$, a value which compares favourably with that measured at low temperatures.

Notice that the gain increases as d is reduced, but the diffraction losses

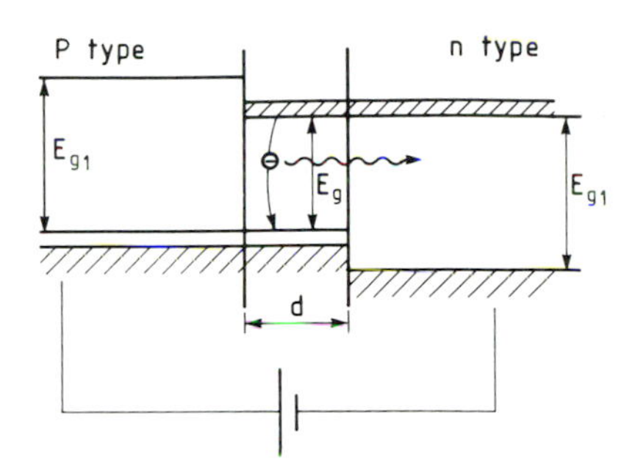

Figure 5.18. A heterojunction semiconductor laser.

increase. However, a proper analysis would require a calculation of the mode structure and its relation to the variation of inversion across the junction. Ideally, we would like to be able to choose the distance, *d*, to obtain the most efficient coupling of the mode structure and electron inversion density. This can be done in the case of heterojunctions, which have now largely replaced simple homojunction lasers. In these devices two (or more) semiconductors with different band gaps are used, as shown in Figure 5.18. The two outer semiconductors can be grown directly onto the central substrate if the crystal lattice dimensions are similar. A good example of such a system is one employing GaAs sandwiched between *p* and *n* type GaAlAs. If $E_{g1} > E_g$, then the effect of a forward bias is to allow the direct injection of electrons and holes into the central region where recombination occurs. Since there is an effective barrier for electrons at the *p*-type interface, and for holes at the *n*-type interface the inversion can be maintained fairly uniformly over the whole of the centre region. Also, *d* is designed into the system rather than being determined by the free carrier diffusion. If the outer semiconductors have a lower refractive index then the whole system acts like a fibre optic waveguide further reducing the diffraction losses. A heterojunction system can be made to operate in CW mode without any cooling. Furthermore, the current requirements are milliamps rather than the amps needed for a simple *p–n* junction.

5.4. Dye lasers

All lasers can be tuned to a certain extent by using intra-cavity elements or selectively filtering the output light. The extent of this tuning is determined, essentially, by the gain profile in relation to the distributed losses. That is, for angular frequencies where

$$\alpha(\omega) - \beta(\omega) > 0 \tag{5.64}$$

β is the total distributed loss, which now, for complete generality, includes frequency dependent terms. Now, the systems discussed so far are tunable over a fairly small range of frequencies. For example, a ruby laser might be tuned over a few nanometres, particularly if the temperature of the crystal can be altered. Over this limited range β may be considered as constant. One class of laser where this does not apply are dye systems.

Dye lasers are broadly tunable systems where the gain profile for one particular dye can extend over 50 nm or more in the visible spectrum. The complete range of tunability, with different dyes, can extend from about $\lambda = 300$ nm to 1500 nm. For this reason, and because of their relatively high efficiency, they have found almost universal applications ih spectroscopy.

The principles of operation of dye lasers are relatively simple, and are based on the fluorescent properties of organic dyes. Most often the dyes are dissolved in liquids, although the amplifying medium can be made into a solid form, using glassy materials, or even acrylic plastic. Lasers have even been constructed, for some research applications, using gelatine for the host material. (The first edible laser!) However, the advantage of using a liquid is that the medium can be circulated, and this has two beneficial effects. Firstly, the cooling can be made more efficient and secondly, the triplet state quenching (discussed later) can be alleviated. The latter is very important in CW systems.

The characteristic feature of organic dyes to fluoresce in visible radiation is brought about by their particular electronic structure, which consists of alternating single and double bonds (conjugate bonds) between the carbon atoms. In any organic compound one can distinguish between σ and π bonds. In σ bonds the orbitals have zero angular momentum about the internuclear axis, whilst π bonds have $\lambda = 1$ (see Section 1.12). An organic compound with double bonds (unsaturated) has three of the four valence electrons in σ bands but the fourth electron is in a π bond. The σ bonds have an electron distribution which is symmetric about the internuclear axis. On the other hand, for π bonds, the internuclear axis lies in a nodal plane (see Figure 5.19). A long chain organic dye can be imagined to consist of a planar structure of σ bonds (and n bonds) with an

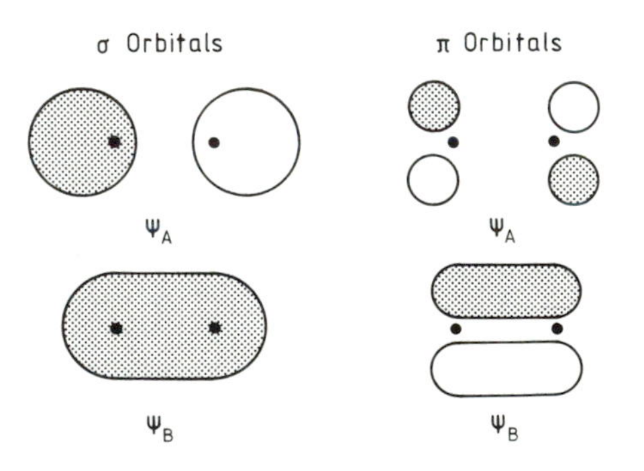

Figure 5.19. Schematic arrangement of σ and π orbitals. Shaded regions are in antiphase to unshaded ones.

electron cloud of π bonds above and below the plane. These electrons are not localized to any one particular atom. The excitation of these relatively free electrons give the dye molecule its ability to respond to long wavelength radiation above 200 nm, whereas saturated compounds have typical electron excitations below 160 nm. In fact, a simple 'free' electron model can be constructed where the electron energies are given by the formula

$$E_n = \frac{h^2 n^2}{8mL^2} \tag{5.65}$$

where n is the quantum number giving the number of antinodes of the wavefunction along the chain, which has a length, L. If there are N π-electrons then the levels are filled to $n = N/2$ (N is even otherwise the system would be an active radical), so the first electronic excited state is formed by taking the electron from the highest occupied orbital and placing it in the first vacant orbital. The energy change, corresponding to the first excited electronic state in the dye, is then

$$\Delta E = \frac{h^2(N+1)}{8mL^2} \tag{5.66}$$

and this corresponds to a wavelength†

$$\lambda_{\text{min}} = \frac{8mcL^2}{h(N+1)} \tag{5.67}$$

For single electron excitations one can construct both singlet and triplet states with the singlet states being lower in energy.

As well as these electronic excitations one must also consider the vibrations and rotations of the molecule as a whole in the manner of Section 1.14. A set of vibrational-rotational levels are built on each intrinsic electronic state. In a dye solution (or in a solid host) the effect of collisions (and electrostatic perturbations) is to smear out these ro-vib states to form a continuous band. Figure 5.20 shows the general band structure of a fluorescent dye such as Rhodamine 6G.

The operation of the dye as a laser medium is shown in Figures 5.20 and 5.21. Optical pumping is used to transfer molecules from the bottom of the S_0 band to the higher levels in S_1 (or even higher S bands). These levels rapidly decay by phonon processes until they fall back to the bottom of the band, or to be more exact, they populate the lower levels according to a Maxwell distribution. These non-radiative processes are extremely rapid and take place in a time of the order

† The change in the standing wave pattern of the electrons in the molecule is transferred into an electromagnetic standing wave!

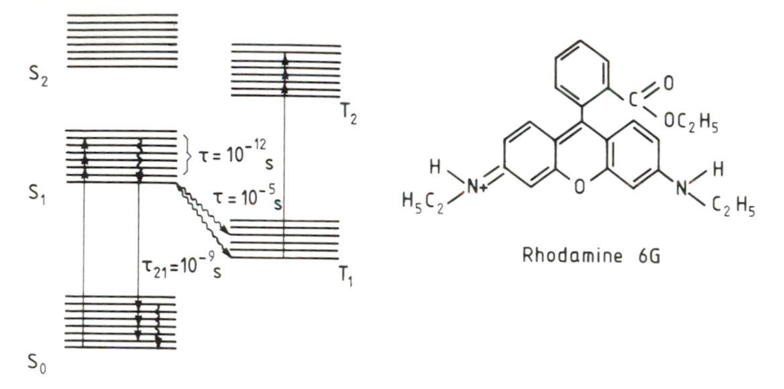

Figure 5.20. General level scheme for a fluorescent dye such as R6G.

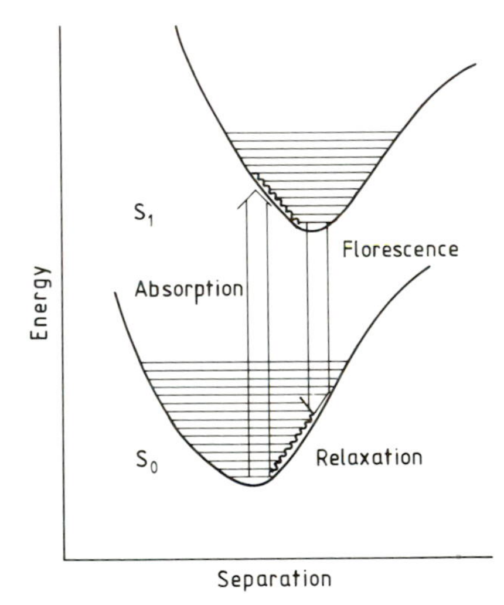

Figure 5.21. Energy diagram illustration of the absorption and emission process.

of 10^{-12} s. Radiative decay ($\tau \sim 10^{-9}$ s) can then take place from the bottom of the S_1 band to the higher levels in the S_0 band, which subsequently relax to the bottom of the S_0 band. Now these radiative absorption and emission processes, between vibrational states, are governed by the Franck-Condon principle which can be understood using Figure 5.21. In its classical form, this principle states that transitions are only possible when there is no change in the internuclear spacings. Thus, electromagnetic transitions can only be drawn vertically on potential energy diagrams like 5.21. In terms of quantum mechanics the principle can be understood by examining the overlap $\int \phi_1 \phi_2 \, dv$ between the vibrational wave functions in the two potentials S_0 and S_1, which essentially determine the allowed transitions. This integral decreases quite rapidly for transitions away from the vertical. A typical dye will have a red shifted emission spectrum which is

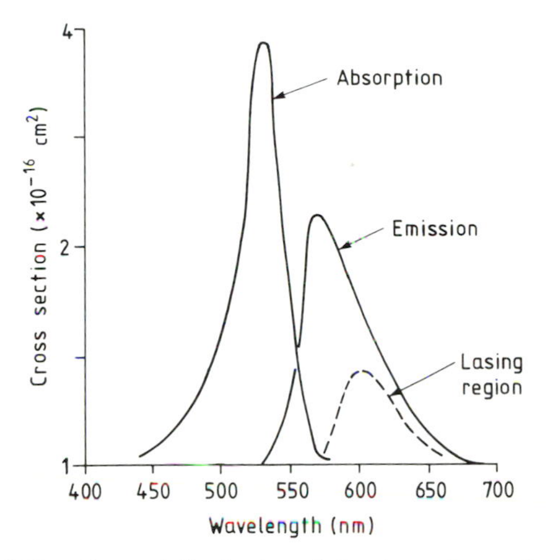

Figure 5.22. Spectral distribution of the absorption and emission in R6G.

approximately the mirror image of the absorption spectrum, as shown in Figure 5.22. (The absorption and emission shapes are often known as the Franck-Condon envelope.) If the two potentials S_0 and S_1 become closer together this will be reflected in the separation of the absorption and emission spectrum. Now we have a four level system which can lase at frequencies where the fluorescent cross section exceeds the absorption cross section.

As previously, these cross sections can be written

$$\sigma_{af} = \frac{g_{af}(\omega)B\hbar\omega}{\eta c} = \frac{\pi c^2 g_{af}(\omega)}{\omega^2\eta^2\tau} \tag{5.68}$$

where g_a and g_f are the absorption and emission lineshapes, which are normalized according to

$$\int g(\omega)\, d\omega = 1 \tag{5.69}$$

An approximate value for the peak cross section can be obtained by assuming that the distribution is a Lorenzian and then using Equation (5.11) together with the known lifetime and widths. Because of the very rapid relaxation between levels with the same electronic structure, the system is effectively homogeneously broadened, but it does not have, as Figure 5.22 shows, a Lorenzian distribution.

If the density of atoms is n_1 in S_0, and n_2 in S_1, then amplification is possible for frequencies where

$$n_2\,\sigma_f(\omega) > n_1\,\sigma_a(\omega) \tag{5.70}$$

However, a more accurate criterion can be written down if we include the other cavity losses, in the manner of Section (3.13). Thus, the laser will oscillate when

$$R\exp(n_2\sigma_f(\omega)L)\exp(-n_1\sigma_a(\omega)L) > 1 \tag{5.71}$$

where L is the length of the active medium and R is the total round trip reflection coefficient ($R = R_1 R_2$). This quantity can be modified to include other losses, as discussed previously. Taking the logarithm and rearranging gives

$$\frac{n_2}{n_1} > \frac{\sigma_a - s/n_0}{\sigma_f + s/n_0} \tag{5.72}$$

Here n_0 is total dye density ($= n_1 + n_2$) and $s = \ln(R/L)$. The pumping required to sustain this inversion is easily calculated for an optically thin sample. Assume a CW system, which is pumped using a fixed frequency laser (a CW gas laser), then

$$\frac{dn_2}{dt} = \frac{-n_2}{\tau} + \frac{n_1 I_p \sigma_p}{\hbar\omega} = 0 \tag{5.73}$$

where σ_p is the absorption cross section at the pump frequency. Hence the minimum pump intensity is

$$I_p = \frac{n_2\hbar\omega_p}{n_1\tau\sigma_p} \tag{5.74}$$

which can be written, using (5.72), as

$$I_p = \frac{(\sigma_a(\omega) - s/n_0)\hbar\omega_p}{(\sigma_f(\omega) + s/n_0)\tau\sigma_p(\omega_p)} \tag{5.75}$$

where ω is the laser frequency and ω_p is the pump frequency.

A serious drawback of any dye laser concerns the loss of molecules by nonradiative transfer to the triplet state, a process which has a lifetime of approximately 10^{-6} s. The fluorescent decay from T_1 to S_0 is spin forbidden ($\Delta S = 1$) and is relatively slow,having a lifetime of the order of 10^{-5} s. Molecules can therefore accumulate in T_1 such that after a few hundred microseconds the population of T_1 will be 10 times that of S_1. This in itself lowers the efficiency because these molecules can no longer take part in the amplification, but a more serious problem arises because transitions are now possible from T_1 to higher triplet states. These are allowed transitions, and, unfortunately, the absorption overlaps the fluorescent emission (as well as lowering the pump efficiency because of absorption). Our Equation (5.75) would need to include the triplet absorption which can be so large as to prevent oscillation. This is not a limitation in a pulsed laser if the flash lamp (or other pumping source) is shorter than, say, 10 μs. In CW systems the build up can be avoided by continuously circulating the dye, thus allowing the molecules which accumulate in T_1 to return to S_0. The volume of dye in the active region has to be changed very rapidly so that this technique is only suitable if this volume is small. This can be achieved if the pumping is accomplished using another laser which can be focused onto a small region of dye. Normally the dye is squirted from a nozzle, rather like water from a tap.

The number and diversity of dye laser systems is quite large, and only a very brief introduction can be given here. Perhaps the simplest type is a flashlamp pumped laser, similar in design to the ruby laser except that the crystal is replaced with a dye cell. Tuning the output can be accomplished in many ways. For example, coarse tuning can be accomplished using a prism, or Lyot filter (a length of birefringent material with polarizers on either side), in the cavity. Rather more precise tuning can be achieved using a reflection grating, for example, an echelle grating, in place of one of the mirrors. Tuning is then carried out by altering the angle of the grating. This latter system is often employed when the dye pumping is done using a pulsed fixed frequency laser (e.g. N_2, excimer or frequency doubled YAG laser). A simple scheme is shown in Figure 5.23.

The effect of these tuning elements is to only allow oscillation over a narrow range of frequencies determined by the response of the element(s). In essence we are taking only a narrow slice of the gain profile represented by Equation (5.75). Even in this small frequency region there will usually be many cavity modes so that single mode operation (single frequency) will not be possible. (Although the system is homogeneously broadened the gain difference between successive modes is small when only coarse tuning elements are present. Also spatial hole burning causes mode competition.) However, if the overall cavity length is reduced to make the mode spacing large, it is possible to achieve single mode operation with a combination of high resolution gratings. An alternative way, with longer cavities, is to use one or more intra-cavity etalons in conjunction with the coarser tuning element. For example, in the case of a single etalon coupled with a grating, the free spectral range of the etalon is chosen so that only a single transmission peak (see Figure 3.10) lies within the tuning envelope of the grating. If there are just a few cavity modes within the transmission of the etalon, then only a single mode at the peak will oscillate. This is similar to our discussion in Section 3.13.

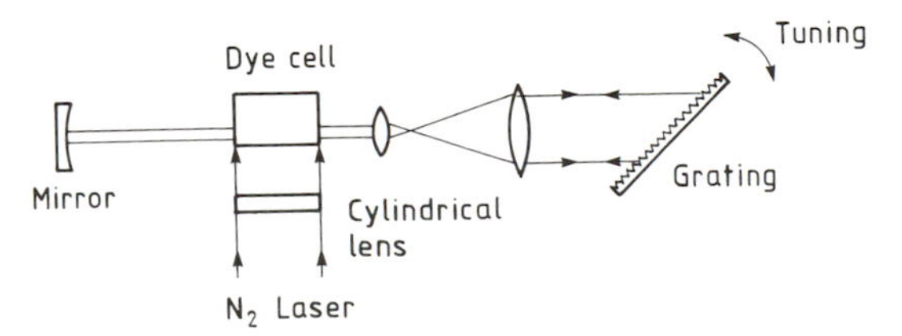

Figure 5.23. A simple N_2 pumped dye laser. Note that the spreading of the beam both lessens any damage to the grating and also increases its resolution.

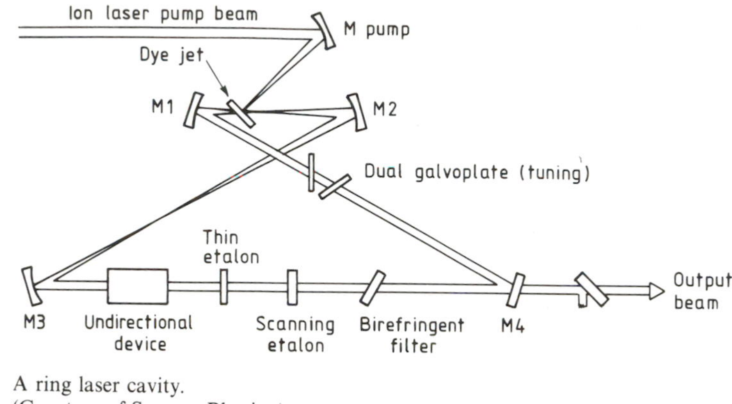

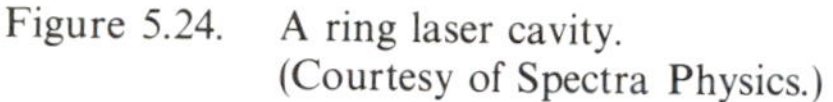
Figure 5.24. A ring laser cavity.
(Courtesy of Spectra Physics.)

The most precise single frequency dye lasers, in terms of linewidth, are CW systems which can be pumped using argon or krypton ion lasers (see next chapter). These are often ring lasers, rather than standing wave lasers, since their power output is higher. A ring laser geometry is shown in Figure 5.24. Instead of forming standing waves the light can circulate indefinitely around the cavity in both directions. Normally, a unidirectional device†, is used to prevent light circulating in one of the directions thus increasing the power in the other direction. The mode separation is now given by the equation

$$n\lambda = L_c \tag{5.76}$$

where L_c is the total length around the cavity. This condition is just that required for the wave to be in phase after travelling around the cavity. Now the laser amplification is truly that of a travelling wave and spatial hole burning is eliminated. With these systems about 1 W of power can be obtained for 6 W pumping power. Coarse tuning is achieved using a birefringent filter, whilst single mode selection is brought about by means of two etalons. Having set the birefringent filter to the approximate wavelength, the laser can be scanned over the free spectral range of the thin etalon, by altering the cavity length using two rotating glass plates while at the same time adjusting the scanning etalon to maintain a fixed mode number. With appropriate changes of dye (and optics) the system can be tuned across the visible spectrum. For the narrowest linewidth and most stable operation the laser can be actively stabilized from a system of external interferometers which sense any change in wavelength outside that allowed by the scanning. Error signals are fed back to the laser cavity to control the cavity length, using either the tuning plates or moving one of the mirrors by means of a piezo crystal.

---

† Based on the Faraday (or magneto-optic) effect. These devices can also be used as modulators.

Bibliography

Allen, L., 1969, *Essentials of Lasers* (Oxford: Pergamon Press).

Barltrop, J.A. and Coyle, J.D., 1975, *Excited States in Organic Chemistry* (London: Wiley).

Driscoll, W.G. and Vaughan, W. (eds), 1978, *Handbook of Optics* (New York: McGraw-Hill).

Firth, W.J. and Harrison, R.G. (eds), 1982, *Lasers—Physics, Systems & Techniques* (Proceedings of the Twenty-Third Scottish Universities Summer School in Physics, Edinburgh).

Levine, A.K. (ed.), 1968, *Lasers*, Vol. 2 (New York: Marcel Dekker).

Levine, A.K. and DeMaria, A.J., 1976, *Lasers*, Vol. 4 (New York: Marcel Dekker).

Weber, M.J. (ed.), 1982, *Handbook of Laser Technology*, Vol. 1 (Florida: CRC Press).

CHAPTER 6

gas lasers

6.0. *Introduction*

In this last chapter, the most common gas laser systems are described, with particular reference to the atomic (molecular) physics. The basic physics of light amplification relevant to gaseous systems has already been discussed, in Chapters 3 and 4, so all that remains is to pick out the details for specific lasers. Indeed, gas lasers are most akin to a system of isolated atoms (they cannot be completely isolated, of course!) and, correspondingly, their distinguishing feature is that the transitions often have narrow linewidths, typically several orders of magnitude less than solid or liquid lasers. Pumping of gaseous systems can be carried out in a variety of ways. Optical pumping is one possibility, particularly for molecular lasers operating in the infra-red, where efficient pumping is possible using another laser of shorter wavelength. However, the most common type of excitation, for lasers in the visible spectrum is using collisions in a gas discharge. Most of this chapter will be devoted to systems which are pumped using this method. Direct electron beam pumping is similar in concept and is briefly mentioned in excimer lasers. It is also possible to use a chemical reaction to produce a population inversion. A simple chemical laser is described at the end of the chapter.

Before embarking on descriptions of particular lasers it is worthwhile reviewing the basic physics involved in gas discharges. This is a subject which could fill a whole text (or more), and its importance in practical gas laser design is paramount. However, we shall be content only to describe the very basic ideas necessary for a complete understanding of the workings of individual lasers. A detailed mathematical description is beyond the scope of this text.

6.1. *Population inversion mechanisms in gas discharges*

There are several types of discharge employed in laser pumping, but all the atomic mechanisms are the same, in that the primary excitation is via electron–atom† collisions within a plasma. The simplest system, suitable for low power

† In this section atom could be replaced by molecule.

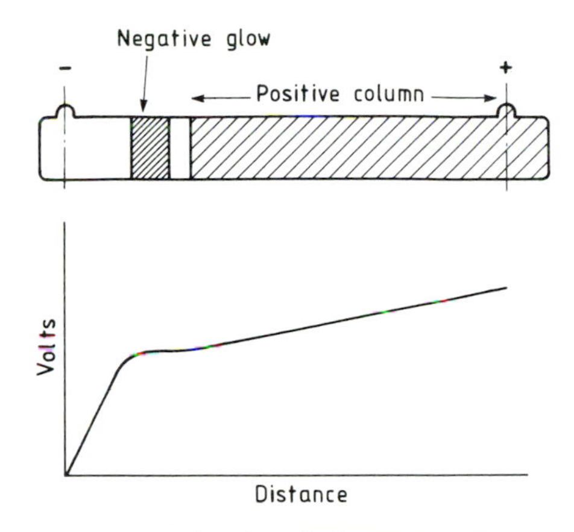

Figure 6.1. Schematic of a glow discharge.

operation, is a d.c. glow discharge. Figure 6.1 shows this arrangement with the electric field along the laser axis. In this cold cathode arrangement, it is usual for the laser amplifier (e.g. He–Ne laser) to be the positive column of the discharge. A similar low pressure system can also be excited by an r.f. field. For the d.c. system, the striking voltage to initiate the discharge will be quite a lot higher than the operating voltage, which for a short tube† might be a few hundred volts at pressure of a few torr. Along the positive column there is a fairly uniform voltage gradient, implying that the plasma is neutral, or, what amounts to the same thing, the number of electrons and positive ions is everywhere equal. (There can be space charge effects at the walls of the plasma tube or at the anode but these can be ignored for the considerations here.) Also, the number of atoms which are ionized is only a very small fraction of the neutral atoms.

The plasma is stable if the number of electrons, produced by electron impact ionization, balances the losses, which are mainly due to capture processes. (Here we are ignoring wall losses which can play an important role in narrow bore laser tubes.) Each electron is undergoing many collisions with a typical time between collisions of the order of 1 ns. Most of the collisions are elastic, where the energy of the electron changes by only a very small fraction, depending on the atomic mass and velocity of the atom, and the angle of scatter. These collisions result in a general heating of the gas. The energy of the electron does not increase until it equals the applied voltage because it is reduced by less frequent inelastic collisions, where the atom becomes ionized or left in an excited state. These inelastic collisions result in much larger losses of energy corresponding to the ionization potential or the excited state energy. A plasma can be characterized by the electron velocity distribution which depends, for a given geometry and gas mixture, on the ratio of the electric field to the total number density of atoms, E/n.

† The critical parameter is the electric field.

For example, a simple glow discharge plasma might have a mean electron energy of 1·5 eV in the anode column. Note that this 'thermal' energy is several orders of magnitude larger than the mean drift energy induced by the electric field ($=\frac{1}{2}mV_d^2$, where $V_d =$ drift velocity). Here the word thermal should be treated with caution since the electron energies are not, in general, described by a Maxwell distribution, although the atomic energies can be. It is, however, fairly common to use the word temperature (in eV) when describing the energies of the electrons. Now clearly there must be a fairly wide distribution of energies, since the plasma could not be sustained, except by collisions with electrons of energy greater than the ionization potential. In fact, it is the high energy tail of the distribution which accounts for these ionizing events. Because the electron distribution depends on the ratio E/n, it therefore follows that there is only one E/n value which can sustain the plasma in equilibrium. Thus we are not free to change the electron distribution to optimize the pumping in low pressure CW systems. Of course, we can alter the geometry, gas mixture or the type of plasma discharge to best suit the requirements.

In some types of laser (e.g. ion lasers) more intense pumping is required, and this can be achieved by having a heated cathode so that a higher current density can be sustained in the plasma. The plasma then has the characteristics of an arc discharge rather than a glow discharge. Another arrangement is the hollow cathode geometry where the amplification is in the cathode dark space. Here the electron energies are higher (see Figure 6.1), thus providing efficient ionization of helium, which can be used to populate metal vapour atoms by thermal energy transfer reactions. Very high current densities can be obtained using pulsed discharges which create plasma conditions that could not be sustained in a d.c. discharge. In these dynamic conditions the electron temperature is higher during the discharge than in the afterglow. Laser action can be obtained in both of these time domains though the detailed mechanisms for producing the inversion are different. Increased electron temperatures can be obtained using a high pressure transverse discharge or TEA (Transverse Electric Atmosphere) geometry. In this arrangement the electric field is applied orthogonal to the lasing direction so that breakdown, in a well defined region, is possible at higher pressures.

The production of a population inversion within the plasma is accomplished in a variety of ways which are best illustrated by the detailed descriptions in the following sections. However, some general observations are worthwhile at this point, and to do this we make reference to the general four level diagrams (Figure 3.2 and Figure 4.2). The primary excitation is by inelastic electron scattering within the plasma leading to excited states of the atom. In some cases (e.g. ion lasers) this is the complete pumping method, but there is no reason, *a priori*, why this should populate the upper laser level in preference to the lower one. Thus, in Figure 3.2, an efficient pumping is one where R_2 is large and R_1 small. A more general method is using a two step process involving the excitation of an atomic species by electron collisions, followed by the resonant (or selective) excitation of the upper laser level in another species by atom–atom collisions. (This is equivalent to the fast relaxation in a doped insulator laser.) A two step process can be made selective if, for example, there is a level in the primary atom close in energy to the upper laser level but there is no level close to the lower laser level. Of course, there will be direct electron excitation of the lower laser level but one way this can be minimized is by having a higher concentration of the 'buffer' gas. It is fairly clear that the problem of optimization is quite complicated but, nevertheless, it is possible to compute these processes fairly accurately if the electron energy distribution in the plasma is known. In fact it is possible to model all the various processes including the ionization as long as the cross sections for all the types of scattering (e.g. electron–atom, atom–atom collisions) are known in the relevant energy range.

Having selective pumping to the upper level is not the complete condition for laser amplification, as was shown in Section 3.2. Thus, in the case of a four level system, Equation (3.27) for the gain can be used to set conditions on the lifetimes and pumping rates necessary to produce a positive gain coefficient. If the degeneracies of the two levels E_1 (lower) and E_2 (upper) are the same then we must have (for CW operation only)

$$R_2 \tau_2 > R_1 \tau_1 \tag{6.1}$$

and

$$\tau_1 < \tau_{21} \tag{6.2}$$

where τ_1 and τ_2 are defined in Section 3.2.

The next sections give brief descriptions of a selection of the most common types of gas laser. Although laser oscillation has been identified in a very large number of neutral† and ionized species, space permits the description of only a few.

6.2. *Helium–neon laser*

This is the archetype of all gas lasers and was the first one to produce visible radiation. In essence, it consists of a helium–neon mixture which is excited in a glow discharge. The laser transitions belong to the excited states of the neon atom which are populated by collisions with excited helium atoms produced by direct electron impacts in the plasma. The principle depends on the existence of excited levels in the neon atom, which are close to the first excited states in helium, and can therefore be populated by resonant neon–helium collisions. In addition, the helium levels are metastable thus ensuring the most efficient transfer because the repopulation of the helium ground state by radiative decay can be ignored.

The level scheme for the laser is shown in Figure 6.2. Here the neon states are

† For example, 730 transitions in 50 elements have been observed.

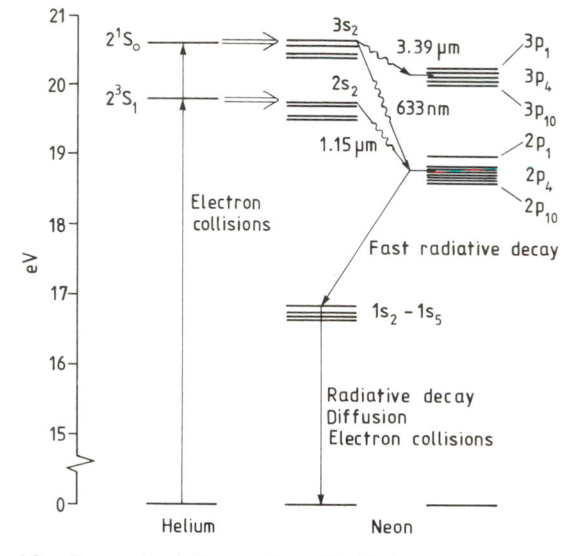

Figure 6.2. Energy level diagram for a He–Ne laser.

labelled by the Paschen notation in recognition of the fact that they cannot be classified by L–S or j–j coupling. (This is true for most of the levels in neon and the heavier rare gases.) The numerical subscripts have no spectroscopic significance and are used only in the way of labels. Thus the ground state is $1s$, and the first excited states, belonging to the $2p^3 3s$ configuration are labelled $1s_2$ to $1s_5$, and those, belong to the $2p^5 3p$ configuration, are labelled $2p_1$ to $2p_{10}$. Above this there are the $2s$ levels belonging to the $2p^5 4s$ configuration, the $3p$ levels, belong to the $2p^5 5s$ configuration, and the $3s$ levels belonging to $2p^5 5s$ configuration. The helium levels are labelled according to the L–S representation which is an excellent approximation in this case. Both the excited 2^3S_1 and the 2^1S_0 levels are metastable. The 2^1S_0 cannot decay by single photon emission since to do so would violate the conservation of angular momentum. This decays by emitting two photons (via the A^2 term of Equation (2.88)) and has a measured lifetime of 19·5 ms. The decay of the 2^3S_1 level by electric dipole radiation is forbidden because of parity violation, and, at first sight, it would seem that magnetic dipole decay would be the obvious mechanism. However, the matrix elements of the dipole moment operator (2.211) vanish unless the two states differ in spin by one unit, or in other words they obey the selection rule $\Delta S = \pm 1$. (This spin flip effect was briefly discussed in Section 2.12). The validity of this selection rule depends on the degree to which S is a good quantum number, which, as we have stated, is certainly true for these states in helium. It turns out that the decay by magnetic dipole radiation is possible when relativistic corrections are added to the simple moment operator, and the calculations yield a decay rate $(1/\tau) = 1{\cdot}27 \times 10^{-4}\,\mathrm{s}^{-1}$, considerably larger than the expected two photon decay. This is rather difficult to measure in the laboratory but in higher z helium-like systems, where lifetimes are much shorter, the calculations based on this mode of decay agree (approximately) with measured values.

Now, by a sheer coincidence, the 2^3S_1 level in helium lies within 0·04 eV of the $2s_2$ neon level, and the 2^3S_0 level in helium lies within 0·05 eV of the $3s_2$ neon level. This makes it possible for the resonant collision reactions

$$\mathrm{He}(2^3S_1)+\mathrm{Ne}(1s_1)\rightarrow \mathrm{He}(1^1S_0)+\mathrm{Ne}(2s_2)+0{\cdot}04\,\mathrm{eV} \tag{6.3}$$

and

$$\mathrm{He}(2^1S_0)+\mathrm{Ne}(1s_1)\rightarrow \mathrm{He}(1^1S_0)+\mathrm{Ne}(3s_2)-0{\cdot}05\,\mathrm{eV} \tag{6.4}$$

to proceed, since the energy difference can be made up by the thermal energy (kT). The collision cross sections are of the order 10^{-17} to $10^{-16}\,\mathrm{cm}^2$. The reverse processes can occur but are not a serious problem since the levels in neon decay rapidly and give rise to the laser transitions. Thus the collisionally excited states are the upper levels (labelled 2 in Figure 3.2 or 3 in Figure 4.2), whilst the lower levels are the $2p$ or $3p$ levels. Now an interesting phenomenon occurs because the $3s_2$ and $2s_2$ levels can also decay by very fast VUV (Vacuum Ultra-Violet) transitions to the ground state, which might be expected to deplete the upper levels and prevent an inversion forming. However, at the gas pressure in the laser, the emitted VUV photons are strongly absorbed, leading to repopulation of the $3s_2$ and $2s_2$ levels. For obvious reasons, this is known as radiative trapping, and, in this case, the net result is to increase the effective radiative lifetime of the two levels to values close to those expected if the VUV mode was not present. (Almost complete trapping.)

The three transitions at 1152 nm, 3391 nm (first gas laser transition) and the familiar red line at 633 nm are the easiest ones to obtain CW oscillations although many others have been made to lase including the $3s_2 \rightarrow 2p_{10}$ green transition at 544 nm. The relative ability to sustain oscillation depends on the small signal gain coefficient in relation to the losses, as we discussed in Section 3.13. For each possible transition, the feeding of the upper and lower level needs to be calculated before Equation (3.27) can be used. This is a non-trivial exercise for a He–Ne system, where there can be several routes, both radiative and non-radiative, for populating (or depopulating) the levels, and, in general, the full calculation would involve the solution of several rate equations. Certain effects can be explained without recourse to algebra. For example, it is not possible to obtain sustained oscillation in the $2p \rightarrow 1s$ transitions simply because the radiative (and other decay modes) of the $1s$ level is much slower than the $2p \rightarrow 1s$ decay, so that atoms simply pile up in the $1s$ states and prevent the formation of a population inversion. This bottleneck has another effect because the fast radiative $2p$–$1s$ decay can be reabsorbed when the population of the neon metastables builds up leading to repopulation of the lower laser levels. However, this radiative trapping is not the major cause of power limitation in He–Ne lasers, as we shall see in the next section.

A simple model can be made to account for the saturation effects observed in the power output of a He–Ne laser. Consider the transitions at 633 nm and 3391 nm, where the upper laser level is the $3s_2$ populated by resonant collisions with He 2^1S metastables. The number of metastable atoms can be calculated by equating the rate at which they are formed to the rate at which they are destroyed by resonant transfer, diffusion to the walls, and electron impact processes. These latter processes include direct de-excitation as well as ionization, and it is assumed, for simplicity, that the ground state of helium is continuously replenished so that it effectively remains constant† with a density, n_0^h. This approach shows how the limiting power output is a result of the saturation of the helium metastables. The rate at which metastables are created is (Rate per unit volume!)

$$R_c = n_0^h n_e \bar{\sigma}_{0m} \bar{v}_e \tag{6.5}$$

where n_e is the electron density and $\bar{\sigma}_{0m} \bar{v}_e$ are averages taken over the electron excitation cross section, σ_{0m}, and velocity, v_e. In terms of the differential cross section, $\sigma(v_e)$, this means

$$n_e \bar{\sigma}_{0m} \bar{v}_e = n_e \int_0^\infty \rho(v_e) \sigma(v_e) v_e \, dv_e \tag{6.6}$$

where $\rho(v_e)$ is the fraction of electrons with energies between v_e and $v_e + dv_e$ and $\sigma(v_e)$ is the cross section per unit velocity interval. Since the metastable is at 20 eV energy only the tail of the electron energy distribution makes any contribution to the integral i.e. $\sigma(v_e) = 0$ for $v_e < v_c$. The destruction rate is

$$R_d = n_m^h n_e \bar{\sigma}_{m0} \bar{v}_e + n_m^h n_0^n \bar{\sigma}_{mn} \bar{v} \tag{6.7}$$

Here $\bar{\sigma}_{m0}$ includes the ionization path, n_0^n is the number density of neon atoms, and $\bar{\sigma}_{mn} \bar{v}$ is an average, similar to (6.6) but now involving the resonant transfer cross section and the atomic velocities. Equating (6.7) and (6.5) at equilibrium gives

$$\frac{n_m^h}{n_0^h} = \frac{k_{1e} i_e}{k_{2e} i_e + n_0^n K_{nm}} \tag{6.8}$$

where the cross sections have been incorporated in the appropriate constants, and i_e is the electron current density. The rate R_2 for the population of the upper

† Otherwise we would need to look at other rate equations, making the balancing much more tedious.

laser transition is then

$$R_2 = n_m^h n_0^n K_{nm} \tag{6.9}$$

which, using (6.8), can be written

$$R_2 = \frac{k_{1e} n_0^h n_0^n i_e}{k_{2e} i_e + n_0^n K_{nm}} \tag{6.10}$$

Population of the lower level is by direct electron excitation for the ground state of neon, and, for the 638 nm transition, by radiative decay of the $2s_2$ level to the $2p_4$ level. For direct electron excitation the rate can be written

$$R_1 = k_{4e} n_0^n i_e \tag{6.11}$$

If we include radiative feeding this term will still have approximately the same form as long as the electron excitation is larger. Combining (6.10), (6.11), and our previous small signal gain coefficient, (3.29), gives the result

$$\alpha = \frac{A n_0^h n_0^n i_e}{\mathrm{B} i_e + C} - D n_0^n i_e \tag{6.12}$$

where the constants now incorporate, in addition to the electron atom excitation cross section, the lifetimes and relative cross sections of Equation (3.28). This equation fits the output power characteristics of the $3s_2 \rightarrow 3p_4$ (3391 nm) transition rather well. In spite of the simplified approach, it is nevertheless true that the output power of the He–Ne laser is limited by saturation of the helium metastables; the primary loss mechanism being due to electron collision ionization. A maximum power of about 100 mW is possible but most commercial systems are considerably less than this. Note the constants in the Equation (6.12) vary when the tube diameter is altered because this affects the average electron energy. If the electron energy loss mechanism is primarily from ambipolar diffusion, then the average electron energy depends (approximately) only on the product of the tube diameter and the pressure. An optimum value for the output power is found when $pd \cong 3{\cdot}5$ torr mm, where p is the total pressure.

A small He–Ne laser generating about 1 mW of CW power is shown in Figure 6.3. In most commercial systems the laser mirrors are integral with the (glass) plasma tube, and, typically, the output mirror would have a reflection coefficient of about 99% at 633 nm. The tube bore would be about 1 or 2 mm diameter and the discharge current between 10 and 50 mA. A typical ratio of helium/neon gas would be 6/1 by volume. The output efficiency of any He–Ne laser is rather poor (between 0·01 and 0·1%) because most of the electron energy which is not lost in elastic collisions goes into ionizing the helium, either directly

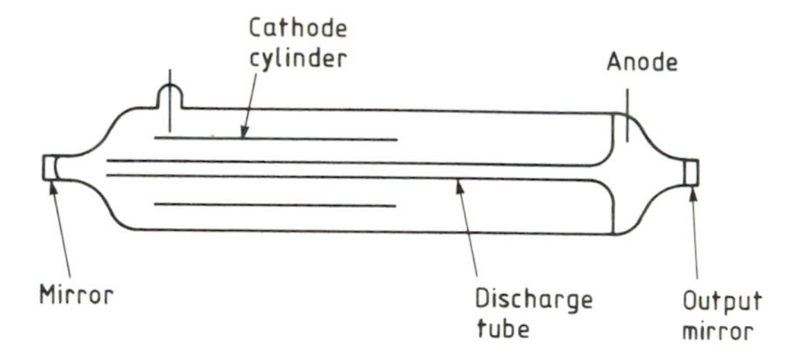

Figure 6.3. A small coaxial He–Ne laser (schematic).

or via the metastable state. At the operating pressure in a glow discharge the gain is dominated by the inhomogeneous effects of Doppler shifts from the thermal velocities of the neon atoms. The broadening will depend on the effective gas temperature. For example, at T = 300 K, Equations (2.180) and (2.183) predict, for neon ($A = 20$), a full width at half height of 1·3 GHz, for the 633 nm transition. The broadening in a laser would be larger than this (2–3 GHz), reflecting the elevated temperature in the plasma. For a laser with a 20 cm mirror separation the mode spacing is 750 MHz so there would be only 3 or 4 modes operating simultaneously. A high pressure waveguide He–Ne laser will have different characteristics because the system is then dominated by the homogeneous effects of pressure broadening. Here only a single mode, near the peak of the gain profile, will propagate (see Section 3.2).

6.3. *Other atomic lasers*

Laser oscillations have been obtained in a large number of neutral atomic species. With the exception of Groups IIIB, IVB, VIB and the actinides, laser transitions have been observed in over 50 elements throughout the periodic table. In all the noble gases (except Ra) many transitions have been made to oscillate on a CW basis using a low pressure gas discharge. Other less favourable transitions have been observed in pulsed discharges. Groups VIIIA and VIIB are a particularly fruitful source of laser transitions, and many of the halides can be made to oscillate on a CW basis. In group IB, copper, silver, and gold contain favourable transitions for pulsed operation. Indeed, copper vapour lasers are finding increasing applications because of their high efficiency for visible radiation. They are discussed separately in Section 6.5.

6.4. *CO_2 laser*

In contrast to the rather poor efficiency of the ubiquitous He–Ne laser, that of the CO_2 can be as large as 30% in some instances. A modestly sized CO_2 system can provide continuous power levels of a few kilowatts, and hence their use in industry, for a variety of applications (such as cutting and welding) is becoming widespread. Pulsed systems incorporating amplifiers can produce terrawatts of power.

Population of the upper laser levels in CO_2 molecules is effected by resonant scattering from excited nitrogen molecules. A glow discharge system would employ a He–N_2–CO_2 mixture, with about 10–15% CO_2 and 10–20% N_2, at a total pressure of 6 to 15 torr. Electron collisions in the plasma produce excited states in N_2 which rapidly transfer their energy to the CO_2. The major purpose of the helium gas is to sustain the plasma discharge (at its optimum value†), but it also plays a role in depopulating the lower laser levels by collisions (relaxation). Since the metastable states in helium are approximately 70 times higher in energy than the lowest vibrational states in nitrogen, the helium has no direct influence in populating the upper laser level. As we found in Section 6.1 it is the tail of the electron energy distribution which produces the metastables in helium (and the ionization!), whereas in the CO_2 laser the bulk of the electrons are involved in the excitation. This is the principle reason for the increased efficiency, apart from the obvious one of operating at a longer wavelength.

Before embarking on a complete description of the laser action the relevance of molecular structure needs to be discussed. Diatomic molecules were considered in Chapter 1 (Section 1.14). N_2 is a symmetric system and, hence, has no permanent dipole moment. The vibrational states (including the rotational levels built on them) cannot decay to other vibrational states with the same electronic structure. In a low pressure discharge these states decay by electron collisions or collisions with other molecules (including the containing wall). This means that the lowest vibrational levels in N_2 are metastable, and these perform much the same function as the He metastables did for the He–Ne laser. In contrast to this, the CO_2 molecule has a fairly complicated structure where there are allowed transitions between the vibrational levels with the same intrinsic electronic structure. The degrees of freedom associated with the collective motion of the molecule are also more complicated than a simple diatomic molecule, and require 4 quantum numbers n_1, n_2^l and n_3. n_1 is the number of quanta associated with a symmetric stretching of the molecule, n_2 with a bending and n_3 with an asymmetric stretch. These are illustrated in Figure 6.4. l is the vibrational angular momentum. For CO_2 (with two ^{16}O or two ^{18}O atoms) even values of n_2, n_3 and $l=0$ are symmetric, while those with an odd value of n_3, an even value of n_2 and $l=0$ are antisymmetric. This means that the rotational level, built on the $00^\circ 1$ vibrational state, has only odd spin members whilst the levels built on the $10^\circ 0$ and $02^\circ 0$ states have only even spin members. Within each group there is a fairly rapid relaxation, at least for pressures encountered in a gas laser. All the members of the same vibrational state will essentially be in thermal equilibrium (the relaxation is slower between vibrational levels!), and the distribution among the

† The mixture and pressure can be adjusted to produce the most efficient pumping.

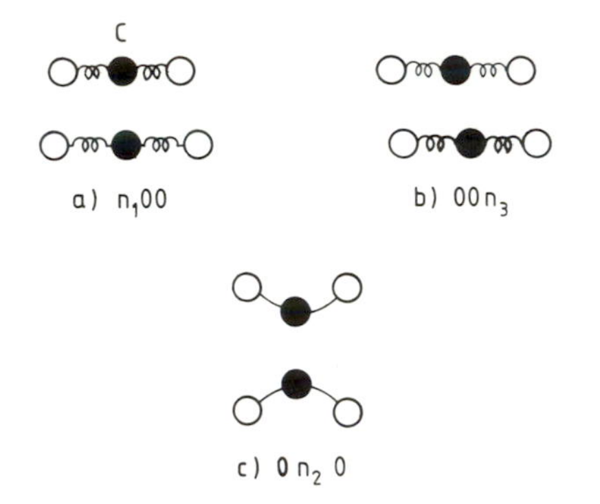

Figure 6.4. Schematic diagram to illustrate the vibrational modes of a CO_2 molecule: (a) symmetric stretching, (b) asymmetric stretching and (c) bending.

levels will be given by Boltzmann statistics. Remembering that the energies are given by Equation (1.28) (ignoring higher order corrections), this is just

$$\frac{N_J}{N_{0v}} = \frac{(2J+1)\exp(-BJ(J+1)/kT_R)}{\sum(2J+1)\exp(-BJ(J+1)/kT_R)} \tag{6.13}$$

where N_{0v} is the total number in the particular vibrational state, v, and B is inversely proportional to the moment of inertia. (This is quite often given in units of cm^{-1} rather than the energy units here.) Notice that the factor $(2J+1)$, which is the total number of states associated with each level, shifts the distribution peak away from the lowest energy member.

Transitions from one set of rotational levels to another must obey the angular momentum conservation rules

$$\begin{aligned} \Delta J = J_{\text{upper}} - J_{\text{lower}} &= +1 \qquad R \text{ branch} \\ &= 0 \qquad\quad Q \text{ branch} \\ &= -1 \qquad P \text{ branch} \end{aligned} \tag{6.14}$$

In most diatomic molecules, the Q branch is excluded because of parity conservation. For CO_2, most of the laser transitions are either P or R because they occur between sets of levels of opposite symmetry. The allowed energies and relative populations can be calculated from the energy formula (1.28), and Equation (6.13). (This is left as an exercise for the reader.) For example, in CO_2 observed transitions in the R branch, of the $00^\circ1$ to $10^\circ0$ band, extend from 10·02 μm ($J_u = 63$ to $J_l = 62$) to 10·39 μm ($J_u = 1$ to $J_l = 0$), and in the P branch, from 10·42 μm ($J_u = 1$ to $J_l = 2$) to 11·18 μm ($J_u = 67$ to $J_e = 68$). Notice the interesting fact that, even if the two vibrational modes have the same total populations, there will be an inversion on the P branch for all values of J

(assuming the constant B is the same for both sets of levels) but not on the R branch. In fact, it can be shown using Equation (6.13) that, even when the ratio $N_{0v'}/N_{0v}$ is slightly less than one, there can still be an inversion for the highest spin transitions of the P branch. This is known as the partial inversion. It must be stressed that a partial inversion can only be obtained when the different internal motions of the molecules (rotations and vibrations) are not in thermal equilibrium, as in a discharge plasma. For a partial inversion the characteristic vibrational temperature, T_v, must be greater than the rotational temperature, T_R. These facts explain why the laser will always predominantly oscillate on the P branch unless constrained to do otherwise. Thus the wavelength for the maximum inversion in the CO_2, $00^\circ 1 \rightarrow 10^\circ 0$ transition, is around 10·6 μm.

A partial energy diagram for the CO_2–N_2 laser is shown in Figure 6.5. The lowest vibrational levels in N_2 are populated via electron collisions in the plasma. Because there is a very close correspondence between these levels and the $00^\circ 1$, $00^\circ 2$, ... levels in CO_2, the cross section for the process

$$\begin{array}{ll} N_2(v = 1) + CO_2(00^\circ 0) \rightarrow N_2(v = 0) + CO_2(00^\circ 1) + 0{\cdot}002\,\text{eV} & \\ (v = 2) \qquad\qquad\qquad\qquad\qquad\qquad (00^\circ 2) & \\ \ldots \qquad\qquad\qquad\qquad\qquad\qquad\quad \ldots & \end{array} \tag{6.15}$$

is large, and leads to a rapid population of the upper laser level. This overlap

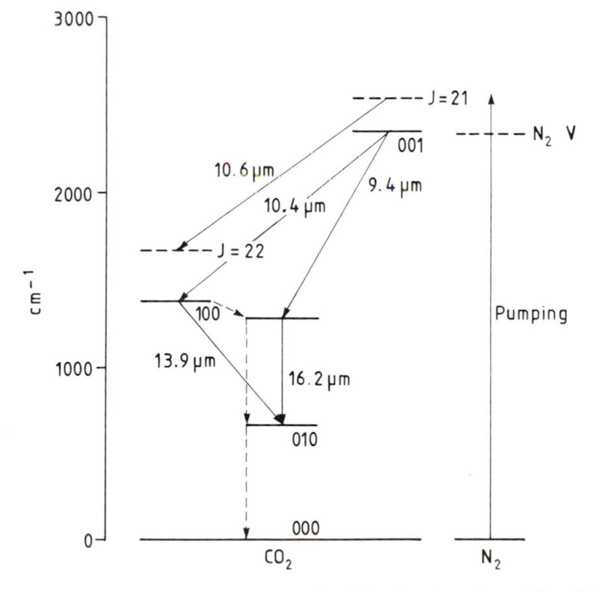

Figure 6.5. Lowest vibrational states in CO_2 showing the origin of the 9·4 μm and 10·4 μm bands. Strong relaxation paths are shown dotted. The position of the energy levels for the strongest transition in the 10·6 μm P branch is also shown.

arises because the restoring force constant (Equation (1.26)) for the asymmetric stretching of CO_2 is very similar to that for stretching of the N_2 molecule. Our figure shows only the lowest levels, but there is also a close correspondence between the $v = 2, 3, \ldots$ vibrational modes of nitrogen and the $00^\circ2$, $00^\circ3, \ldots$ vibrations of carbon dioxide. The pattern of levels repeats itself and gives rise to laser transitions of the same energy, though they contribute less to the overall efficiency. The two most important transitions are between the $00^\circ1$ and $10^\circ0$ vibrational states (10·4 μm band), and the $00^\circ1$ and $02^\circ0$ states (9·4 μm band). It is usual to label the band by the wavelength of radiation which would arise if the two vibrational states decayed directly to one another. Hence, the P branch of the 10·4 μm band extends above this wavelength, peaking at 10·6 μm, and the R branch gives wavelengths less than 10·4 μm.

The lower laser levels are depleted by relaxation processes. Symmetric stretching modes can share their energy with the 2 phonon bending modes, which subsequently relax within the $0n_2^l0$ manifold. The latter vibrational states are then converted into translational energy by collisions with the helium. All of these processes are more rapid than the spontaneous emission, and it is usual to ignore any radiative effects (apart from the stimulated emission!), when calculating the inversion. The main reason for this is that the lifetimes for spontaneous decay vary as λ^{-3} (Equation (2.192)) so they become much larger at these infra-red wavelengths. On the other hand, the stimulated emission cross section is smaller than most visible transitions in a gas, being about 10^{-18} cm^2. For appreciable gain it is necessary to have a fairly large population inversion of CO_2 molecules and, by the same token, most of the molecules will contribute one photon by stimulated emission. For an optimally designed plasma (one where the E/n ratio gives the best pumping), about 55% of the energy put into the discharge can be converted to energy in the upper laser level. Since the quantum efficiency of the laser is about 40% (corresponding to nearly one photon per excited CO_2 molecule) the overall efficiency can be 20% or more. Some of the discharge energy is lost in direct heating ($\sim$15%) (from elastic electron scattering), and some in electronic excitation of the nitrogen ($\sim$30%). A negligible amount is lost in ionization though this process is vital in sustaining the plasma!

Consider now the output characteristics of the radiation if the laser is operating on a particular band. Figure 6.6 shows the relative gain for the P and R branch of the $00^\circ1$ to $10^\circ0$ band, where each line represents a particular transition and is numbered by the spin of the final level. Since $P(22)$ has the highest gain it will start to oscillate and its inversion will decrease relative to the other levels. Now we need to look at the rate at which the upper, $J = 21$, level is replenished and the lower, $J = 22$, level is depleted. This amounts to the same thing as saying how rapidly the rotational levels relax among themselves in the $00^\circ1$ and $10^\circ0$ manifolds. The process could be modelled using coupled rate equations but it can be seen that, for relaxation more rapid than the *effective* lifetime for decay, the different rotational levels will maintain the thermal (Boltzmann) distribution. In CW operation this is exactly what happens and the laser operates on a single

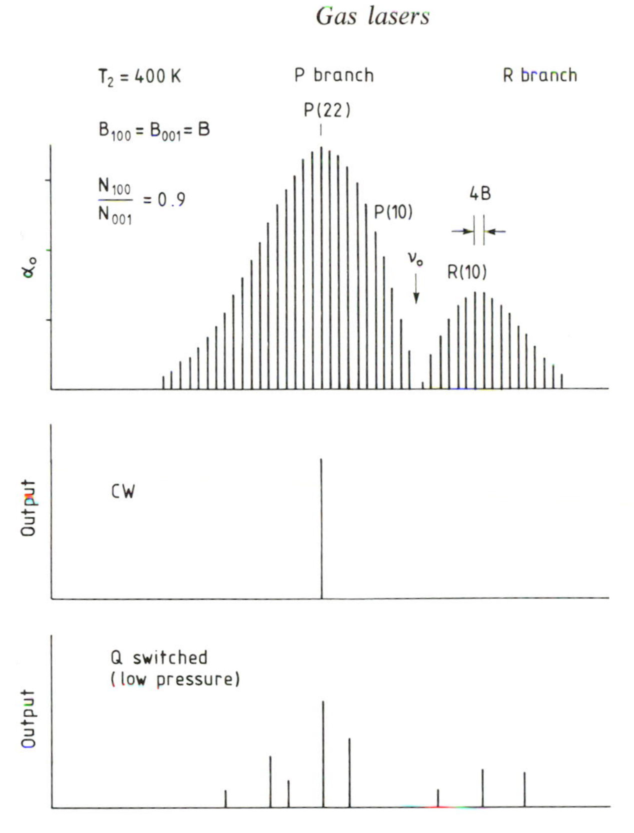

Figure 6.6. The small signal gain of a CO_2 laser (top) at $T_R = 400$ K with $N_{100}/N_{001} = 0{\cdot}9$, for the $10{\cdot}4\,\mu$m band. Shown below is the expected output for CW and *Q*-switched operation.

transition. This would normally be the one with the most gain, that is *P*(22), but oscillation on others can be achieved with frequency selective devices in the cavity. In a *Q*-switch operation we are demanding a much shorter lifetime for stimulated emission and then the laser will oscillate on other transitions, as shown in Figure 6.5. However, if the system is at higher pressure, for example, in a TEA laser, the relaxation is more rapid and the population inversion is maintained highest on a single transition.

Within this single wavelength the gain profile will be broadened inhomogeneously by Doppler shifts, and homogeneously by collisional processes (pressure broadening). At this wavelength the Doppler width is quite small, for example, it would be approximately 50 MHz at an effective plasma temperature of 400 K. Even for a laser with a cavity length of 200 cm†, $c/2L = 75$ MHz, and only a single mode will propagate. The homogeneous broadening is, comparatively, quite large even at low pressures (> 10 MHz) so the propagating wave essentially burns out the whole of the Doppler gain profile. This is even

† Modern high pressure systems tend to be smaller.

more true when cross-relaxation is considered (see Section 2.9), or where the homogeneous broadening is increased, as in a higher pressure system. It is important in all these high power CW systems to maintain oscillations on a single transverse mode (TEM_{00}) where the active medium can be arranged so that a maximum number of molecules are contributing to the output.

Although CO_2 is the most important molecular laser there are many others. Light diatomic molecules like CN, CO, HF, ... or triatomic ones like N_2O, CO_2, CS_2, ... can be made to oscillate at wavelengths in the 2–20 μm range. Molecules such as NH_3, H_2O, CH_3, OH, and a host of others can produce stimulated emission in the far infra-red region. These lasers are usually optically pumped by means of another shorter wavelength laser, for example, a CO_2 system. The reason for this is that, as the density of states increases, the ability to achieve selective excitation, in a gas discharge, diminishes. Lasers based on the electronic excitation in molecules will, of course, operate at shorter wavelengths. These are illustrated in the next sections.

6.5. *Molecular nitrogen laser*

The nitrogen gas laser provides a simple and cheap method of producing intense pulses of near UV radiation, suitable for a variety of applications including the pumping of a pulsed dye laser. The relevant molecular excitations are shown in Figure 6.7. It is usual in molecular physics to label the lowest electronic levels (or manifold) with the symbol *X*, and the increasing excitations of the same multiplicity with the letters A, B, C, D, ... Excitations of different multiplicity are prefixed by *a*, *b*, *c*, *d*, ... in ascending order of energy. However, in N_2 the naming is reversed and the levels of different multiplicity are prefixed *A*, *B*, *C*, *D*, ... The other symbols describing the states were discussed in Chapter 1 (Section 1.12).

Two transitions, *C* to *B*, giving a band with a wavelength centred at around 357 nm† and *B* to *C*, giving several bands from 748 nm to 969 nm, can be made to lase. Note that *A* is metastable (τ is of the order of seconds) because its decay would involve a change of spin multiplicity. The lifetime of *C* is 40 ns whilst that of *B* is 10 μs. These lifetimes mean that CW operation on the UV band is not possible even in the event of super intense pumping because the $B \rightarrow X$ transition would become a bottleneck, unless some scheme for rapid relaxation of the levels was devised.

Pulsed operation using a rapid high discharge current is used to populate the *C* manifold and produce an inversion between *C* and *B*. Two requirements are essential in this three level operation. Firstly, the rate of pumping must exceed the

† This is the $v_i = 0$ to $v_f = 0$ transition but there is also a narrow band from the $v_i = 0$ to $v_f = 1$ vibrational manifolds.

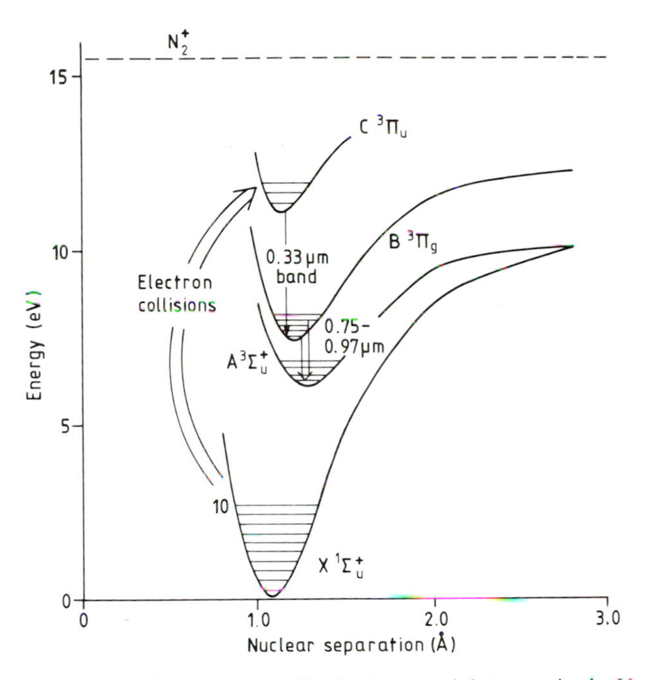

Figure 6.7. Potential energy curves for the lowest triplet states in the N_2 molecule showing the origin of the laser transitions.

spontaneous emission rate, or

$$\frac{dn_c}{dt} > \frac{n_c}{\tau_c} \tag{6.16}$$

where τ_c is the total lifetime of the state c. This requirement is certainly satisfied if a very rapid electronic discharge is used. Secondly, both the rates of population of the initial and final states involve the tail of the electron energy distribution in the plasma, and we therefore require

$$\sigma_{x \to C} > \sigma_{x \to B} \tag{6.17}$$

where $\sigma_{x \to C}$ is the cross section for electron scattering leading to C and $\sigma_{x \to B}$ is that to B. Selective pumping is not, therefore, programmed by using a buffer gas, but nevertheless this inequality is obeyed. The reason for this concerns the Franck-Condon principle which was discussed in Section 5.4. This principle can be equally applied to the excitations of states via the Coulomb field of the scattered electron. The mean separation of the nitrogen atoms in X is 1·094 Å, whilst in C and B it is 1·148 and 1·212 Å respectively. The overlap between the two vibrational wavefunctions is therefore largest for the X to C transition.

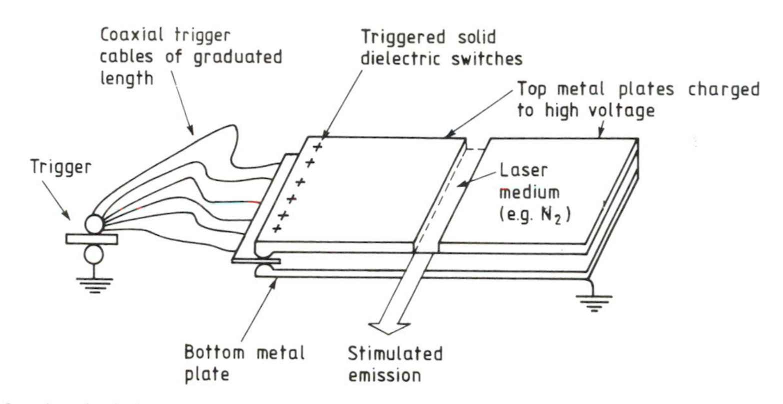

Figure 6.8. A pulsed nitrogen laser employing a Blumlein generator. The discharge is between the top two metal plates and travels at the speed of light to match the stimulated emission. (After J.D. Shipman, *J. Appl. Phys. Lett.* **10** (1967) 3.)

A scheme for a pulsed nitrogen laser, to produce UV radiation, is shown in Figure 6.8. This is a TEA arrangement using a Blumlein generator. Electrical breakdown of the gas is effected by the rapid discharging of one of the high voltage plates. The main problems in such a system are ensuring that the discharge is fast enough and uniform along the length of the spark channel. Although a population inversion can only be maintained for a very short time, the cross section is large and hence the gain is also large. It turns out that there is little point in having mirrors† particularly when using the arrangement of Figure 6.8. A pulse about 4 ns long with (typically) about 4 mJ of energy is produced in a cavity 100 cm long.

Other molecular lasers employing electronic excitations include I_2, H_2, Na_2, Te_2, and excimer lasers. In particular the rare gas halide lasers can be pumped in an electrical discharge. Both systems are sources of UV radiation and are briefly described.

6.6. Excimer lasers

Excimer lasers are different in concept to normal molecular lasers because the lowest energy state of the molecules is not bound. They are divided into two groups according to whether the molecule is formed from two rare gas atoms, or from a rare gas atom coupled to a halide. (Actually the word excimer is a misnomer when referring to any molecule with non-identical atoms since it is derived from the two words, excited dimer).

Consider the general molecular potentials, shown in Figure 6.9, for the production of an excimer AB*. Its formation could be from the collision of an

† In transverse discharges which do not travel along the cavity a single mirror can be used to ensure that 50% of the photons are not wasted!

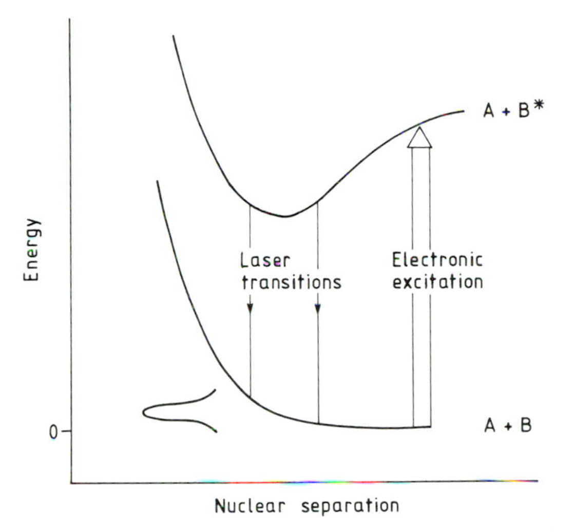

Figure 6.9. Potential energy curves to show the formation of the excimer AB*.

excited atom A* with B, or the collision of a positive and negative ion A^+ and B^-. The characteristic attribute of an excimer is concerned with the changes in potential energy as the internuclear separation is altered. In the lowest electronic configuration the potential is always repulsive (or at least so shallow that it can be thermally dissociated), whilst there are higher electronic configurations which exhibit a distinct minimum energy at one separation. The reason for this behaviour concerns the nature of the valence electron orbitals as a function of the nuclear separation (see Section 1.12). In the excited states there are valence orbitals which decrease in energy as the nuclear separation is reduced. These are the bonding orbitals discussed previously. For example, the polar system H_2 has two electrons in the $1s\sigma$ orbital giving the singlet Σ_g ground state. Since for both electrons there is a decrease in energy as the nuclear separation is reduced, we expect the system to be the most highly bound of all. Physically, the increase in binding energy arises because the valence electrons can share the excess nuclear charge in the most beneficial way. Thus, the electrons in the H_2 molecule show a distinct preference for localizing between the two nuclei where the extra attractive energy more than offsets any increase in repulsive energy.

To illustrate the formation of an excimer, consider the noble gases Ar, Kr and Xe. The ground state electronic configuration for these gases corresponds to the complete filling of the $3p$, $4p$ and $5p$ shells respectively. These are strongly bound states where the residual Coulomb interactions between electrons are small. In their ground state, they are monatomic and do not form chemical bonds because they have no valence electrons. Apart from the small attractive contribution from van der Waals' forces (at longer ranges), they exhibit the general repulsive potential shown in Figure 6.9. At separations close to the point where the electron clouds overlap ($= r_1 + r_2$ where r_1 and r_2 are the effective hard

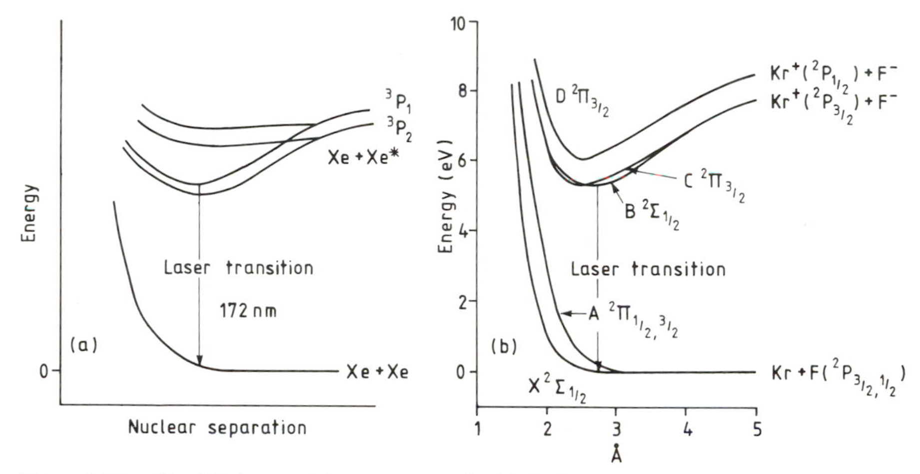

Figure 6.10. Simplified potential energy curve for (a) the lowest states of Xe_2 and (b) the lowest levels in KrF.

sphere radii), the slope of the potential curve becomes extremely large. Now consider a covalent molecule formed from a ground state atom and an excited atom, for example, Xe + Xe*. The lowest excited state of Xe corresponds to the promotion of a single electron from the 5*p* shell into the 6*s* shell, and therefore has the configuration... $5s5p^5 6s$. The lowest excited states are represented by a single hole in the 5*p* shell coupled to an electron in the 6*s* shell, which, in *LS* coupling, gives rise to the triplet levels $^3P_2$, $^3P_1$ and $^3P_0$, and the singlet level $^1P_1$. The lowest energy molecular levels are the $^3\Sigma_u^+$ and $^1\Sigma_u^+$ states, which arise from the two lowest atomic states, $^3P_2$ and $^3P_1$. A general potential energy diagram is shown in Figure 5.10. Notice that above these two levels are two further states, $^3\Sigma_g^+$ and $^1\Sigma_g^+$, which have only a shallow potential minimum.

In the two bound excited states the molecule can decay radiatively with a lifetime of about 10 ns† ($A \sim 10^8\,s^{-1}$) for the $^1\Sigma_u^+ \rightarrow {}^1\Sigma_g^+$ transition, and a larger value for the $^3\Sigma_u^+ \rightarrow {}^1\Sigma_g^+$ transition which is spin forbidden. The electromagnetic transition ($\lambda = 172\,nm$) therefore corresponds to the single valence electron falling back into the hole in the 5*p* shell of the Xe core. Once in the lower state the two xenon atoms separate rapidly in about the time it takes for one vibrational period, or approximately 0·1 ps. (Remember that it is the vibrational-rotational levels in the upper state which are decaying). Because this lifetime is so short the final state energy is uncertain, and thus the radiation is broadened by about 5 nm for Xe_2 (generally in the range 0·3 to 30 nm). This broadening allows the laser to be tuned over a wider range than atomic gas lasers where the usual limit is set by the Doppler-gain profile. An equivalent description of this tuning can be made on a classical basis using the Franck-Condon principle. In the upper bound state the

† Corresponding to a fully allowed dipole transition.

atom is vibrating and can only decay by vertical transitions on the energy diagram. Thus, if two lines are drawn vertically from the turning points of the vibrational motion they will intersect the unbound ground state potential at two points, which represents the approximate tuning range.

The peak cross section, from (5.11), is simply

$$\sigma = \frac{\pi \lambda\!\!\!^{-2}}{\tau_{21}\gamma_t} \tag{6.18}$$

where γ_t is the half width of the homogeneously broadened profile. Typical values for excimers range from $\sigma = 3 \times 10^{-18}\,\mathrm{cm}^2$ to $3 \times 10^{-16}\,\mathrm{cm}^2$ depending mostly on the broadening, which in turn is related to the molecular potentials, that is the steepness of the potential in the lower state at the dissociation radius. Although the lifetime is short the cross section is rather small because the broadening is large. On the other hand, this is more than compensated for by having a system where the lower level is depleted almost instantaneously, thus allowing for a large inversion to be produced as long as the upper level can be populated efficiently. The rather high losses in the system (from the broadband absorption of the UV light), coupled with the smallish cross section, mean that the input power is certainly too large to sustain on a CW basis. This can be calculated quite simply, since the minimum input power (per unit volume) to sustain a density, n_2, in the excimer, is

$$P = \frac{n_2 \hbar\omega}{\tau_2} \tag{6.19}$$

even if all the energy goes into the photons. Since we also have

$$\alpha = \beta = n_2 \sigma_{21} \tag{6.20}$$

at threshold, then

$$P = \frac{\beta\hbar\omega}{\sigma_{21}\tau_{21}} = \frac{\beta\hbar\gamma_t c}{\pi\lambda\!\!\!^{-3}} \tag{6.21}$$

is the minimum power density (note the general scaling as $1/\lambda^3$). For example, β is typically $5 \times 10^{-3}\,\mathrm{cm}^{-1}$, which, for $\tau = 10^{-8}\,\mathrm{s}$, and $\sigma_{21} = 10^{-18}\,\mathrm{cm}^2$, gives a power density of $190\,\mathrm{kW\,cm}^{-3}$. Of course, the actual power required will be somewhat larger than this.

Rare gas excimers are pumped using pulsed electron beams. In one common arrangement a thin wall tube (50 μm) containing Xe is made the anode of a vacuum diode. A cylindrical cathode surrounds the anode and is connected via a suitable line to a high voltage generator (e.g. Marx generator giving 500 kV pulses). The inside surface of the cathode has sharp projections on it which enhance and spread the electron emission during the vacuum discharge. In this way a very intense pulse of electrons, lasting for a few hundred nanoseconds, can be produced. These have sufficient energy to penetrate the thin anode wall and deposit most of their energy in the gas. Almost all of this is taken up with ionizing collisions, producing Xe^+ ions which then combine to give Xe_2^+. Following this the Xe_2^+ ions undergo several reactions before finally ending up in the Xe_2^* state. It turns out that the pumping is very efficient with something like 40% of the overall electron energy being converted to radiation in the excimer band. The reason for this is that there is no strong, single pathway during the collisions for the system to lose energy, except via the excimer, which can then only decay radiatively. To understand this we would be aware that above the excimer energy there are many intersecting potential energy curves. In a collision the two atoms travel along one curve but where it intersects another, with a steeper negative slope, they can move apart with a lower energy. The final reaction leading to the excimer is a three body one and hence the laser requires a high operating pressure. Yet again this is a manifestation of the Franck-Condon principle because the excited states alter their electronic configuration only at a fixed separation, but this time it is a relaxation process rather than a radiative one. Within any one electronic configuration the vibration-rotational levels can lose energy by collisions and we therefore expect all the levels up to about kT above the excimer zero point energy to be populated.

Although rare gas excimers are reasonably efficient, their power output tends to saturate as the input is increased. This lack of scaling is thought to be due to photo-ionization of the $^3\Sigma_u^+$ state which tends to thermally equilibrate with the $^1\Sigma_g^+$ state as the electron current density is increased. Because the transition to the ground state is spin forbidden the photo-ionization cross section exceeds the stimulated emission cross section for the $^3\Sigma_u^+$ state.

Photo-ionization of the excimer is not a problem in rare gas halide lasers because the wavelength of light is too long to cause ionization. Indeed, these excimers have largely replaced rare gas systems as a general means of producing UV light†, except where the very shortest wavelengths are required. (The wavelengths of the bands range from 175 nm for ArCl to 351 nm, for XeFl. Kr_2 and Ar_2 produce output wavelengths of 146 and 126 nm respectively). An excited rare gas atom is similar in nature to an alkali metal and hence the bonding between it and a halide atom might be expected to be fairly strong. Figure 6.10 shows the molecular energies for KrF. Production of the excimer in electron beam pumping is fairly complicated and will not be discussed here. In fact, these systems are still being actively researched and some details of the mechanisms are not fully understood. A large *e*-beam pumped system can produce pulses of several hundred Joules energy, with durations of about 600 ns, whereas about 10 J is the limit for a large rare gas excimer.

† For example, the small rare gas halide lasers can be used as an efficient means of pumping an organic dye.

Another advantage of rare gas halide systems is that gas discharge pumping has been developed so that small systems, with modest output energies (per pulse), are available to most laboratories. The pumping is by means of a fast transverse discharge (10 ns long), which at high gas pressure, requires some form of pre-ionization, or resistive ballasting, if the discharge energy is not to be channelled into a single arc of small transverse area. This is often done using UV radiation from a number of point-to-plane gaps in series with the main discharge. An arc in these gaps provides the necessary pre-ionizing radiation whilst at the same time completes the circuit for the transverse discharge.

6.7. *Copper vapour laser*

Copper vapour lasers provide an efficient means of producing visible green radiation, albeit in the form of short pulses at high repetition rates. The radiation has many applications, ranging from the pumping of laser dyes to underwater communications.

A level scheme for atomic copper is shown in Figure 6.11. The five lowest levels belong to configurations involving the $3d$, $4s$ and $4p$ atomic orbits which are fairly close together in this part of the atomic table. (See Figure 4.1 for the binding energies of the last electrons.) The ground state is $^2S_{1/2}$ and has the configuration $3d^{10}4s$, whilst the first two excited levels, which are designated $^2D_{3/2}$ and $^2D_{5/2}$, correspond to the promotion of a single electron from the $3d$ shell to the $4s$ shell. These states can be approximately represented as a single hole in the $3d$ shell. Above this are two further excited states belonging to the $3d^{10}4p$ configuration with assignments $^2P_{1/2}$ and $^2P_{3/2}$. Because the $3d$ shell is full these can be described

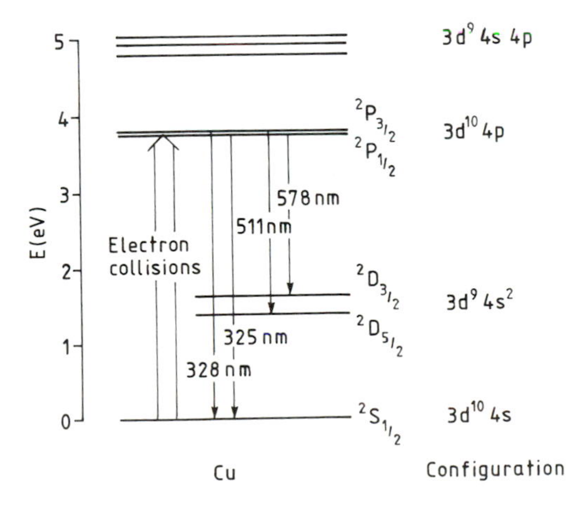

Figure 6.11. Energy diagram for Cu.

approximately by wavefunctions of a single p electron†. Notice that the $3d^9 4s$ and $3d^{10} 4s^2$ configurations have even parity whilst the $3d^{10} 4p$ configuration has odd parity. Fully allowed transitions ($E1$ radiation) are possible between the P levels and the $^2S_{1/2}$ ground state, as shown. These correspond to single electron transitions between a $4p$ and a $4s$ orbit, and have radiative lifetimes of about 10 ns. On the other hand the radiative transitions from the P to D levels, although fully allowed by selection rules, will have a longer lifetime ($\tau \cong 100$ ns) since the transition has less energy and also, it involves a more complicated 'rearrangement' of electronic structure. However, the states are not described by pure configurations and a small amount of mixing allows the transitions to proceed by $E1$ radiation. For example, the D states may also contain small amounts of the even parity configuration, $3d^{10} 4s^2$. $E1$ transitions from the $D_{3/2}$ and $D_{5/2}$ levels to the ground state are strictly forbidden on parity grounds as well as being ΔJ-forbidden, for the $D_{5/2}$, transition and ΔL-forbidden, for both transitions.

At first sight it would therefore seem impossible that laser oscillations on the 511 and 578 nm lines could be expected. Certainly the isomeric nature of the D levels makes CW oscillations impossible, but how can a reasonable inversion be maintained when the competing return path to the ground state is so rapid? The answer is that, at the vapour pressure of copper in the discharge, the UV radiation is almost completely trapped.

Pumping of the upper laser level is by means of a rapid longitudinal discharge through copper vapour in a noble gas buffer. Because the vapour pressure of copper is inherently low, the plasma temperature needs to be about 1500°C before there is a sufficient density of copper atoms. Population of the upper level is by direct electron excitation from the collision of plasma electrons and copper atoms. The lower laser levels will also be populated, but the electron cross section for this process is so much smaller that it is not a serious problem. This large difference in cross section is because the excitation to the D states involves the promotion of a 'core' electron from the $3d$ shell. A rapid discharge therefore creates an inversion which can be sustained for 30–50 ns. Surprisingly, this is much shorter than the radiative lifetime of the laser transition; the quenching in this case being due to collisions of electrons and copper atoms in the excited P states. The efficiency of the laser can be as high as 1% and typical pulse energies, for a (reasonably sized) commercial system, might be about 5 mJ at pulse repetition frequencies of a few kHz. Peak powers are therefore 100 kW ($\tau = 50$ ns), and average powers, 20 W, at 4 kHz.

6.8. Ion lasers

The previous sections have discussed neutral atomic and molecular lasers

† These are not the hydrogen wavefunctions since screening has to be included (see Section 1.5).

with particular emphasis on the most commonly occurring systems. Now we briefly review lasers which employ transitions in ionized species. The discussion will refer mainly to single ionized atoms which can be produced in a gas discharge plasma of modest power.

Noble gas ion lasers

In a gas discharge with a fairly high current density the positive ions can become excited by further electron collisions. For noble gas ions there are sets of levels at a high excitation (>20 eV for argon) which can be excited in a gas discharge. The lifetimes of these states are such that CW oscillation is possible for many transitions. To illustrate this, we look at the argon ion laser, since this is the most common system, and is used extensively for pumping a CW dye laser.

Figure 6.12 shows the relevant energy diagram for the argon ion. The important characteristic of this scheme is that the upper laser levels and the ground states have odd parity, whilst the lower laser levels have even parity. Decay of the upper laser level to the ground states is forbidden whilst the lower laser level can decay rapidly, particularly since this is an energetic transition in the UV spectrum. (The lifetimes according to simple calculations are expected to be about 0·3 ns). Now the lifetimes of the most favourable laser transitions are of the order 10 ns, for example, the $^4D^0$ state has a measured total lifetime of 1·5 ns. This means that the criteria (6.1) and (6.2) can be met even when the pumping rate R_1 is greater than R_2, as is most certainly the case for an argon ion laser. Radiative trapping is not a limiting problem except at the very highest powers because the density of argon ions is low. It is also further reduced because the plasma tube is

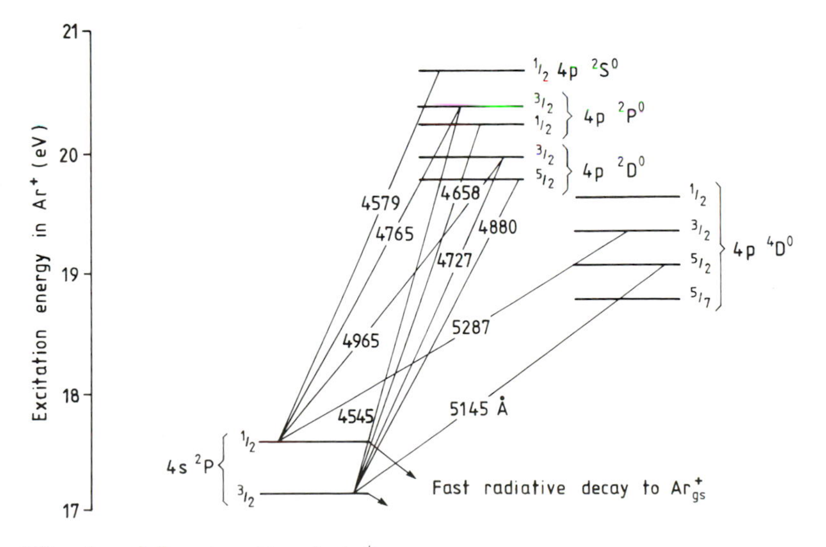

Figure 6.12. $4p \rightarrow 4s$ laser transitions in Ar^+.

narrow in order to achieve the necessary conditions of high current and high electron temperature. The current density has to be high in order to achieve a sufficient population in the upper laser level. As discussed earlier, it is the high energy tail of the electron distribution which causes the ionization. For this, the electron energy needs to be greater than 16·7 eV, and, even if the peak of the distribution is around 5 eV, the number of electrons of this energy is small. For example the energy of the electrons falls off approximately as $\exp(-E_e/kT)$ where E_e is the electron energy and T is the 'electron' temperature. To further excite the argon ion requires an electron of energy greater than ~20 eV, so only a further small fraction of the plasma electrons can contribute. Notice that the direct production of the excited ions requires an electron of energy >35 eV and can therefore be disregarded in view of the exponential fall in numbers. Because of the two step nature of the population production we expect the variation of output power to be proportional to the square of the electron current density. Figure 6.13 shows that this is obeyed in a practical laser system until the point where double ionization occurs.

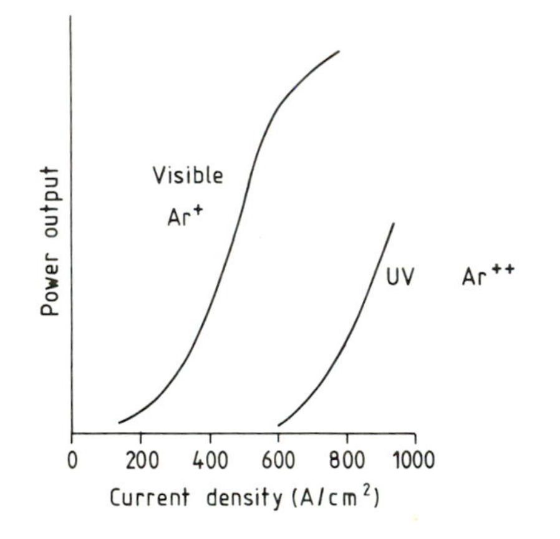

Figure 6.13. Relative output power for Ar^+ and Ar^{++} transitions.

The argon ion laser works because of the favourable lifetime ratios of the transitions coupled with the fact that there is no serious limitation to increasing the number density in the excited state until the threshold value has been exceeded. The optical losses in the system can be made small so that oscillations are possible in spite of the fact that the inversion cannot be high. (The number density of ions in the excited state can only be a small fraction of the number density of argon atoms.) By the same token the efficiency of the system is extremely low. Most of the electron energy goes into ionizing collisions producing argon ions which ultimately lose their energy at the plasma wall. Even for those ions which are further excited the quantum efficiency is only 7%. (This is

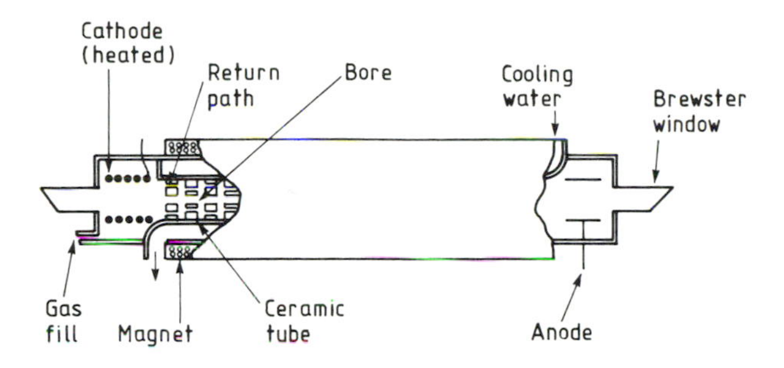

Figure 6.14. An argon ion laser (schematic).

just the ratio of the maximum energy extracted as stimulated emission to that necessary to produce excited ions. It therefore represents the maximum possible efficiency.) Thus we expect the overall efficiency to be low, and figures between 0·05 and 0·2% are typical for a CW system.

An argon ion laser giving about 5 W of continuous power is shown in Figure 6.14. The plasma is produced in a narrow bore tube, 2·5 mm diameter by 0·5 m long, containing argon at a pressure of about 0·5 torr. The cathode is heated to give the necessary large current densities and, in this condition, there is a fairly uniform drop in voltage unlike the characteristics of a glow discharge. Besides requiring a large current density it is also clearly important to have a high electron temperature. The current density can be increased by having a higher pressure, but only at the expense of lowering the electron temperature which depends on the mean free path of the electron between collisions. However, if the tube diameter is reduced the electron temperature can be increased for a given pressure. This amounts to the same thing as saying that the plasma will support a higher voltage for a given current when the tube diameter is small. This is because most of the energy in the electrons (and the argon ions) is lost at the walls rather than in making more argon ions. The electrons therefore need to do more work in order to sustain the discharge and hence there is a larger voltage across the plasma. It is also found that an axial magnetic field (about 0·15 T) along the discharge tube gives a higher efficiency, although the reasons for this are not, at present, fully understood. At the ends of the plasma tube are windows, inclined at Brewster's angle, which allow the passage of linearly polarized light without suffering significant scattering losses. A typical value of the operating current is about 30 A at 400 V, and the gain under these conditions (for the strongest visible transition), might be about 1% per cm. If the current in the tube is increased further then the gain begins to saturate, mainly because of the formation of Ar^{++} ions. Indeed, UV transitions in the doubly ionized argon can be made to oscillate as shown in Figure 6.13. The gain for these transitions will be significantly less than the prominent visible ones, and the output mirror will need a lower transmission for optimum power output (c.f. Equation (3.162)). Most lasers have mirrors in the form of a multiple layer of dielectric coatings which are designed to

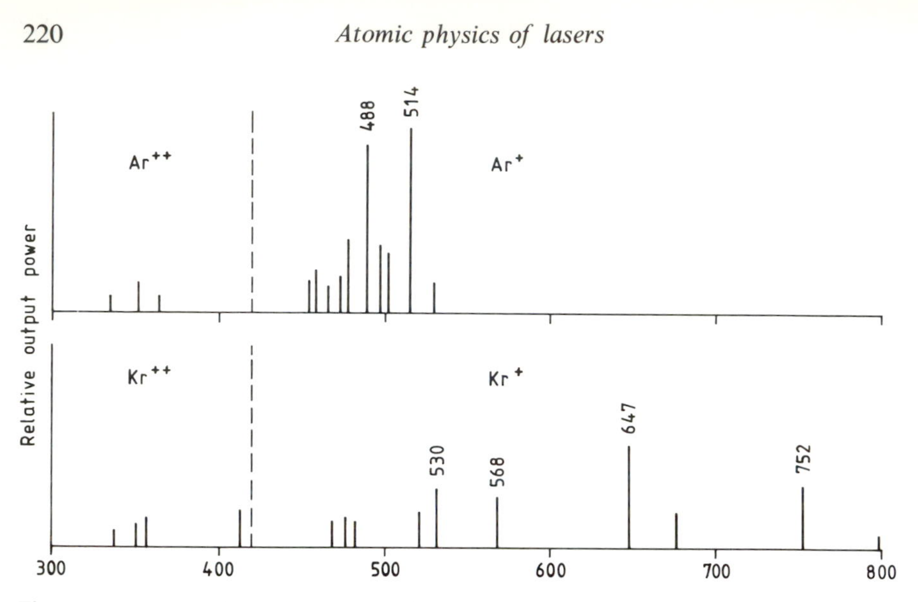

Figure 6.15. Lines observed in argon and krypton ion lasers.

give the correct reflection/transmission characteristics only over a fairly narrow range of wavelengths. In these circumstances, the mirrors need to be changed when the laser is made to operate on the UV transitions.

The output wavelengths of argon and krypton ion lasers are shown in Figure 6.15. Selection of a single wavelength can be achieved with an intracavity prism. Ion lasers have fairly long cavities so there will be several modes within the gain profile, which is essentially dominated by Doppler broadening. Single mode operation can be achieved by introducing an etalon into the cavity. Most lasers are designed to operate only in the TEM_{00} mode, a characteristic which is usually ensured by the necessity of having a narrow plasma tube. However, the curvature of the mirror(s) needs to be chosen carefully in line with the analysis of Section 3.10.

Cadmium ion laser

A laser which is similar in some respects to the noble gas ion lasers is the cadmium ion laser. Discharges in a helium buffer gas containing a trace of cadmium give oscillations on transitions in the excited cadmium ion. The levels belong to configurations of the type $4d^{10}5s$, $4d^{10}5p$ and $4d^9 5s^2$, and are shown in Figure 6.16. Transitions from the $4d^{10}5p$ levels to the ground state are rapid with lifetimes of the order of a few nanoseconds. The laser transitions at 441 and 325 nm are between the $4d^9 5s^2$ and the $4d^{10}5p$ levels. Although these are allowed by selection rules they involve two electron jumps and might therefore be expected to have very long lifetimes. However, the wavefunctions describing these states are not made from pure configurations so that there is some mixing of the $4d^9 5s^2$ and $4d^{10}5d$ configurations which have the same parity. Only a small

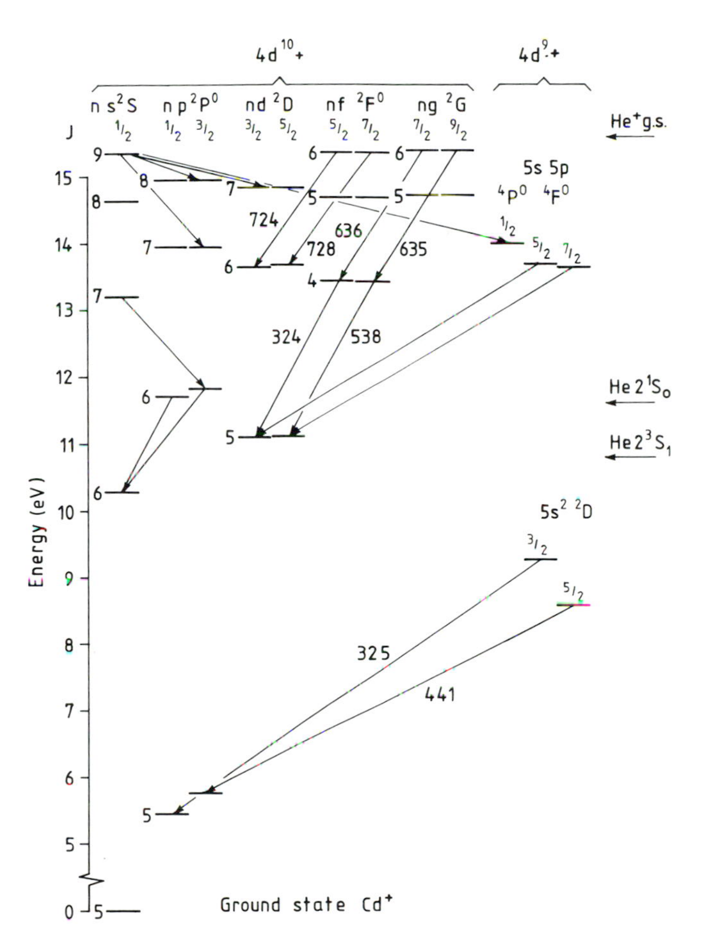

Figure 6.16. Laser transitions observed in Cd$^+$.

amount of mixing is required to reduce the lifetime to its value of 500 ns. As in the case of an argon ion laser the lifetime ratios are very favourable for oscillation on the upper transitions even if $R_2 < R_1$.

In addition to the direct electron excitation, the upper laser level can be populated by Penning ionization. In this process the 2^3S_1 helium metastables have enough energy to produce excited cadmium ions during collisions with cadmium atoms. The process which leads to the population of the $^2D_{5/2}$ level can be written

$$\text{He}(2^3S_1) + \text{Cd}_{g.s.} \rightarrow \text{He}_{g.s.} + \text{Cd}^+(^2D_{5/2}) + e + 2{\cdot}24\,\text{eV} \qquad (6.22)$$

Because the final energy is also shared with an electron the process has a much broader resonance than a simple two body scattering, and hence the excess

2·24 eV can be carried off by the electron. Selective population of the upper laser levels is therefore ensured by the process.

Zinc ions have a similar electronic structure to cadmium except that they now involve the 3*d*, 4*s* and 4*p* shells. Strong oscillations are observed at 589, 621 and 747 nm corresponding to transitions between the $3d^9 4s^2\ {}^2D$ and the $3d^{10} 4p\ {}^2P$ levels.

The Cd^+ laser is a convenient and easily constructed source of UV radiation albeit at fairly low power. Typical discharge currents of 150 mA in a tube about 8·5 mm diameter, 300 mm long would give 10 mW of radiation. Powers of a few hundred milliwatts have been achieved in larger systems although the output tends to flatten off with increasing discharge current. This behaviour is due to saturation of the helium metastables, and is similar to the effects observed in a He–Ne laser.

A different excitation method is possible for populating the higher energy levels in Cd^+ and several other elemental ions including Ag^+, Au^+, Zn^+, Hg^+ and I^+. This is by charge exchange with He^+, or in some cases Ne^+, represented by the general reaction

$$\mathrm{He}^+_{g.s.} + B \rightarrow \mathrm{He}_{g.s.} + B^{+*} \qquad (6.23)$$

where B^{+*} is the laser ion upper level. This is known as thermal charge exchange and has energy defects in the range from 0·1 to 2 eV. The process is usually more specific than Penning ionization because there is no longer a third body (i.e. the electron) to remove the excess energy. For example, in Cd^+ the levels belonging to the configuration $4d^{10}6g$ (${}^2G$ term) are at about the same excitation above the ground state as the He^+ is above the $He(1\,{}^1S_0)$ ground state. A strong laser transition, giving radiation of 635 and 636 nm, is possible between these two upper levels and the two levels belonging to the ${}^2F$ term of the $4d^{10}4s$ configuration.

Usually, the discharge is of the hollow cathode type. Only in the region of the cathode dark space is the density of He^+ high enough for the pumping rate, R_2, to be sufficient.

6.9. Chemical lasers

A laser employing a chemical reaction to provide the input energy is an attractive alternative to one pumped using a gas discharge, particularly in industrial applications where the latter may require large electrical installations. For example, in the chemical reaction

$$H_2 + F_2 \rightarrow 2HF \qquad (6.24)$$

about $1{\cdot}3 \times 10^4$ J of energy is produced for every gram of 'fuel' consumed.

Assuming that a system with 10% efficiency can be made (which is modest compared with a good CO_2 laser) then a simple CW laser with kilowatts of output, and using nothing more than bottled gas, could be envisaged. Fairly high power chemical lasers have been constructed but, as yet, they do not compete favourably with electrical discharge pumped CO_2 lasers.

The general principles of operation depend on the population of rotation-vibration levels following a chemical reaction. For example, hydrogen and fluorine can 'burn' in the two step reaction

$$F + H_2 \rightarrow HF + H + \Delta E_1 \tag{6.25}$$

and

$$H + F_2 \rightarrow HF + F + \Delta E_2 \tag{6.26}$$

This is a chain reaction which can be initiated by UV radiation, electron beams or by an electrical discharge. The first reaction by itself can be obtained by means of a discharge in a SF_6/H_2 mixture. In (6.25) ΔE is 31·7 kcal/mole and this means that vibrational states (based on the *g.s.* electronic configuration), up to $v = 3$ in HF can be populated, whilst in (6.26) ΔE_2 is 97·9 kcal/mole allowing population of vibrational-rotational levels up to $v = 9$. The population of the various levels depends on the rates of the chemical reaction as well as the various relaxation rates. Ideally, one would like the reaction to be 'resonant' so that only one vibrational state is populated but this is not borne out in practice. For example, it is found in chemiluminescence measurements of the first reaction (6.25) that the $v = 3$ and $v = 1$ states are about equally populated. The population among the different rotational levels is quite complicated. Indeeed, the strongest transition is the $P(5)$ member of the $v = 2$ to $v = 1$ band with a wavelength of 2·795 μm. One of the major difficulties in these chemical lasers is caused by the build up of *g.s.* HF which strongly absorbs the laser radiation. (This effect is due to the regular (harmonic) spacing of the vibrational-rotational levels). In CW lasers the gaseous constituents are 'burnt' in a supersonic jet so that the HF is continuously removed. This arrangement also serves to remove the large quantities of extraneous heat generated during the reaction.

6.10. *Gas dynamic lasers*

Chemical lasers are closely akin to gas dynamic lasers, which are now briefly described. Again, the laser transitions are between the vibration-rotation levels of a molecule in its lowest electronic configuration. Population of the excited states is by direct heating, a process which always results in a negative inversion between any two levels, as long as thermal equilibrium is established. However, the real subtlety comes in the next stage of the process which involves the rapid

Table 6.1. Relaxation rates for excited CO_2 and N_2 molecules

Reaction	$k[\mathrm{cm^3\,mole^{-1}\,s^{-1}}]$
$CO_2(001)+CO_2 \rightarrow CO_2(000)+CO_2$	$2{\cdot}4\times10^9$
$CO_2(001)+H_2O \rightarrow CO_2(000)+H_2O$	$1{\cdot}8\times10^{11}$
$CO_2(010)+CO_2 \rightarrow CO_2(000)+CO_2$	$4{\cdot}8\times10^9$
$CO_2(010)+H_2O \rightarrow CO_2(000)+H_2O$	$6{\cdot}0\times10^{12}$
$CO_2(010)+N_2 \rightarrow CO_2(000)+N_2$	$6{\cdot}0\times10^8$
$N_2(v=1)+N_2 \rightarrow N_2+N_2$	$<6{\cdot}0\times10^3$
$N_2(v=1)+H_2O \rightarrow N_2+H_2O$	$1{\cdot}2\times10^9$

adiabatic cooling of the gas by expansion in a series of (or a single) supersonic nozzles. The inversion is created by the different rates of relaxation of the molecules in the upper and lower laser levels, and is best illustrated by the CO_2 gas dynamic laser. Here the 100 relaxation is much more rapid than the 001 relaxation so that the population of the former diminishes more rapidly than the latter as the jet leaves the nozzle region, and at some point the number density in the upper state starts to exceed that in the lower state. Mirrors can be placed transversely to the jet at a point where the inversion is largest.

In essence the laser exploits the differing cross sections for energy transfer during the scattering of molecules with 001 and 100 vibrations. Usually the CO_2 is mixed with a N_2 gas buffer which performs a similar role to that in a discharge CO_2 laser. The relaxation between the 001 levels and the N_2 $(v=1)$ levels is rapid and leads to an equilibrium between the two modes. Because the nitrogen vibration to translation relaxation is slow, the N_2 acts as an effective energy storage for populating the upper laser transition during the expansion. Hence it is usual to have a much higher partial pressure of N_2 in the system. The relevant relaxation rates for the various processes (vibration to translation only) are given in Table 6.1. Included in the table are the figures for H_2O which is often added to the mixture. This considerably speeds up the loss of 001 molecules but also has an even greater effect on 100 molecules. Note that the important 100 to 010 relaxation is not included because this is so fast that the two can be considered to be in equilibrium. In fact the laser exploits both the features which make the discharge CO_2 laser so effective, namely a rapid relaxation of the lower levels and a selective population of the upper level via excited N_2 molecules.

In a practical laser the gas mixture might be about 91% N_2, 9% CO_2 and 1% H_2O heated to 1400 K. At this temperature the Boltzmann equation would predict about 1·5% of the CO_2 molecules in the 001 state and 4% in the 100 state. On expanding the mixture to produce a stream at a velocity of Mach 4 the resulting rapid cooling produces a dramatic change in the populations. At 4 cm from the nozzle the population of the 100 vibration is less than 0·1% of the total whilst the upper level is reduced only by a factor of one half. Of course, a lot of the stored energy comes from the population of the $v=1$ N_2 molecule.

Bibliography

Fenstermacher, C. (ed.), 1976, *High Energy Short Pulse Lasers* (New York: IAA Publications).

Firth, W.J. and Harrison, R.G. (eds), 1982, *Lasers—Physics, Systems & Techniques* (Proceedings of the Twenty-Third Scottish Universities Summer School in Physics, Edinburgh).

Levine, A.K. (ed.), 1968, *Lasers*, Vol. 2 (New York: Marcel Dekker).

Verdeyen, J.J., 1981, *Laser Electronics* (New Jersey: Prentice Hall).

Weber, M.J. (ed.), 1982, *Handbook of Laser Technology*, Vol. 2 (Florida: CRC Press).

INDEX